Photocatalytic Water and Wastewater Treatment

Edited by
Alireza Barzagan

School of Environment
College of Engineering
University of Tehran, Iran

Published by   IWA Publishing
Unit 104–105, Export Building
1 Clove Crescent
London E14 2BA, UK
Telephone: +44 (0)20 7654 5500
Fax: +44 (0)20 7654 5555
Email: publications@iwap.co.uk
Web: www.iwapublishing.com

First published 2022
© 2022 IWA Publishing

Apart from any fair dealing for the purposes of research or private study, or criticism or review, as permitted under the UK Copyright, Designs and Patents Act (1998), no part of this publication may be reproduced, stored or transmitted in any form or by any means, without the prior permission in writing of the publisher, or, in the case of photographic reproduction, in accordance with the terms of licenses issued by the Copyright Licensing Agency in the UK, or in accordance with the terms of licenses issued by the appropriate reproduction rights organization outside the UK. Enquiries concerning reproduction outside the terms stated here should be sent to IWA Publishing at the address printed above.

The publisher makes no representation, express or implied, with regard to the accuracy of the information contained in this book and cannot accept any legal responsibility or liability for errors or omissions that may be made.

Disclaimer
The information provided and the opinions given in this publication are not necessarily those of IWA and should not be acted upon without independent consideration and professional advice. IWA and the Editors and Authors will not accept responsibility for any loss or damage suffered by any person acting or refraining from acting upon any material contained in this publication.

British Library Cataloguing in Publication Data
A CIP catalogue record for this book is available from the British Library

ISBN: 9781789061925 (Paperback)
ISBN: 9781789061932 (eBook)
ISBN: 9781789061949 (ePUB)

This eBook was made Open Access in March 2022

© 2022 The Editors

This is an Open Access book distributed under the terms of the Creative Commons Attribution Licence (CC BY-NC-ND 4.0), which permits copying and redistribution for non-commercial purposes with no derivatives, provided the original work is properly cited (https://creativecommons.org/licenses/by-nc-nd/4.0/). This does not affect the rights licensed or assigned from any third party in this book.

Photocatalytic Water and Wastewater Treatment

Contents

Foreword ix

About the Editor xiii

Chapter 1
An introduction to photocatalysis *1*
Mohammad Reza Boskabadi, Vincent Rogé, Alireza Bazargan,
Hamed Sargazi and Emanuele Barborini

1.1 Principles of Photocatalysis 1
1.2 Mechanism and Characterization of the Photocatalytic Process. 3
    1.2.1 Photocatalytic mechanism 3
    1.2.2 Lifetime and mobility characterization of photo-excited
            charge carriers 5
    1.2.3 Photocatalysis characterization by organic dye degradation . . . 9
1.3 Homogeneous and Heterogeneous Photocatalytic Processes. 10
    1.3.1 Homogeneous example: photo-Fenton reaction. 10
    1.3.2 Heterogeneous photocatalytic reaction examples 11
1.4 Applications of Photocatalytic Processes 13
    1.4.1 Water splitting 13
    1.4.2 Solar energy 13
    1.4.3 Reduction of carbon dioxide. 13
    1.4.4 Water and wastewater treatment 14
1.5 Photocatalytic Materials. 17
    1.5.1 Semiconductors 17
    1.5.2 Heterostructures 18
    1.5.3 Z-Scheme heterostructures. 20
1.6 Operational Parameters Affecting the Photocatalytic Process 22
    1.6.1 Effect of pH. 23
    1.6.2 Temperature 23
    1.6.3 Presence of oxidants 24
    1.6.4 Pollutant concentration. 24

vi **Photocatalytic Water and Wastewater Treatment**

     1.6.5   Catalyst loading 26
     1.6.6   Light intensity and wavelength 26
1.7   Conclusion 28
References 28

Chapter 2
Metal organic frameworks for photocatalytic water treatment **37**
Mohammadreza Beydaghdari, Mehrdad Asgari and
Fahimeh Hooriabad Saboor

2.1   Introduction 37
2.2   Basis of Photocatalytic Degradation 38
2.3   Metal-Organic Frameworks as a Photocatalyst 40
     2.3.1   Photocatalytic degradation of dyes using MOFs 43
     2.3.2   Photocatalytic removal of pharmaceuticals and personal
             care products (PPCPs) using MOFs. 49
     2.3.3   Photocatalytic removal of heavy metals using MOFs 52
     2.3.4   Photocatalytic degradation of pesticides and herbicides
             using MOFs. 54
     2.3.5   Photocatalytic degradation of other organic pollutants
             using MOFs. 55
2.4   Parameters Affecting the Photocatalytic Performance 56
     2.4.1   Light intensity and wavelength 57
     2.4.2   Photocatalyst loading 57
     2.4.3   Dissolved oxygen 57
     2.4.4   The effect of pH 57
     2.4.5   Initial concentration of contaminants. 58
2.5   Conclusions and Outlook 58
References 59

Chapter 3
Photocatalytic reactor types and configurations **73**
Meysam Shaghaghi, Hamed Sargazi, Alireza Bazargan and
Marianna Bellardita

3.1   Introduction 73
3.2   Basic Components in the Design of Photocatalytic Reactors. 75
     3.2.1   Characteristics of light 75
     3.2.2   Catalyst 79
     3.2.3   Process type 83
3.3   Photoreactor Configurations 85
     3.3.1   Bed reactors 85
     3.3.2   Fluidized bed reactors. 87
     3.3.3   Thin-film reactors 88
     3.3.4   Membrane reactors 91
     3.3.5   Cascade reactors 91

Contents vii

3.3.6 Spinning disk reactors. 93
3.3.7 Rotating photocatalytic reactors 96
3.3.8 Taylor vortex photoreactor 97
3.3.9 Optical fiber reactors. 98
3.4 Concluding Remarks 100
References 104

Chapter 4
Landfill leachate treatment using photocatalytic methods **111**
Hamed Sargazi, Alireza Bazargan, Meysam Shaghaghi and
Mika Sillanpää

4.1 Introduction 111
4.2 Landfill Leachate Characteristics 112
4.3 Landfill Leachate Treatment 117
4.4 Homogeneous Photocatalytic Processes for Leachate Treatment 118
4.4.1 Combination of photo-Fenton methods with other
treatment options. 119
4.4.2 Operational parameters 121
4.5 Heterogeneous Photocatalytic Process for Leachate Treatment 121
4.5.1 Modification of nanocatalysts 123
4.5.2 Operational, structural and environmental parameters 124
4.5.3 Combination of the heterogeneous photocatalytic
method with other treatment methods 127
4.6 Conclusion 127
References 129

Chapter 5
Life cycle assessment of solar photocatalytic wastewater treatment. *135*
Mohammadreza Hajbabaie, Hossein Nematollahi,
Ka Leung Lam and Alireza Bazargan

5.1 Introduction 135
5.2 The Basic Steps of LCA. 137
5.2.1 Goal and scope. 137
5.2.2 Life cycle inventory 138
5.2.3 Life cycle impact assessment 139
5.2.4 Interpretation. 141
5.3 LCA for Wastewater Treatment and Photocatalysis 141
5.4 Materials and Methods of the Current Study. 144
5.4.1 Goal and scope of the current study 145
5.4.2 Inventory analysis of the current study. 145
5.4.3 Life cycle impact assessment of the current study 145
5.5 Results and Discussion 149
5.6 Conclusion 155
References 155

viii Photocatalytic Water and Wastewater Treatment

Chapter 6
Analysis of patents in photocatalytic water and wastewater treatment. Part I – photocatalytic materials *159*
Ali Zebardasti, Mohamad Hosein Nikfar,
Mohammad G. Dekamin, Emad Sanei, Itzel Marquez and
Alireza Bazargan

6.1   Introduction 159
6.2   Patent Registration Document Analysis 161
    6.2.1   The main steps of patent analysis 162
6.3   Patent Analysis for Photocatalytic Materials in Water and
    Wastewater Treatment 167
    6.3.1   The registrants of patents in the field of photocatalytic
        materials for water and wastewater treatment 170
    6.3.2   Key extracted concepts 174
6.4   Conclusion 176
References 180

Chapter 7
Analysis of patents in photocatalytic water and wastewater treatment. Part II – solar energy and nanotechnology *183*
Ali Zebardasti, Mohamad Hosein Nikfar, Danilo H. S. Santos,
Lucas Meili, Mohammad G. Dekamin and Alireza Bazargan

7.1   Introduction to Solar Photocatalytic Patents 184
7.2   Analysis of Solar Photocatalytic Patents 186
7.3   Nanotechnology in Photocatalytic Treatment of Water
    and Wastewater 196
7.4   Concluding Remarks 204
References 205

Index 209

Foreword

The challenges which result from water stress and inadequate water-related hygiene have taken on an increasing urgency in the past one or two decades, inextricably linked as they are to the water-food-energy nexus, in the midst of which global climate change now has the potential to create the perfect storm. A recent UN report has warned that 'overcoming the crisis in water and sanitation is one of the greatest human development challenges of the early 21st century'.

It is estimated that about one in every six people in the world today lack proper access to clean drinking water, and double that number lack basic sanitation. The awful result is that 2.2 million deaths per year are related to water poverty, many of these children. This humanitarian tragedy is only forecast to grow worse, with more than half of the world's population forecast to face chronic or critical water shortages by 2050, thereby further limiting food supplies and, in turn, limiting economic and social development. There is no doubt that ensuring an adequate water supply to underwrite a sustainable future presents a challenge to engineers of the first magnitude.

Measures which are adopted to meet the needs of individuals must be sustainable if they are going to be truly effective. The United Nations Conference on Sustainable Development in Rio de Janeiro in 2012 created a vision of 'commitment to sustainable development and to ensuring the promotion of an economically, socially and environmentally sustainable future for our planet and for present and future generations'. Water and sanitation are prominent in the document:

We recognize that water is at the core of sustainable development as it is closely linked to a number of key global challenges. We therefore reiterate the importance of integrating water into sustainable development, and underline the critical importance of water and sanitation within the three dimensions of sustainable development...We stress the need to adopt measures to significantly reduce water pollution and increase water quality, significantly improve waste-water treatment and water efficiency and reduce water losses. In order to achieve this, we stress the need for international assistance and cooperation.

The sustainable solution to these challenges will not lie in new and massive water reservoirs, wells, pipelines, or long-distance river transfers, nor will it solely lie in large-scale salty water desalination. A more sustainable development and use of finite freshwater resources is required, which in turn will bring into existence more sophisticated and resource-circular technologies. For example, regions which become more water-stressed as a result of climate change might choose to rely more and more on brackish ground waters or seawater as a primary source of water, this being energy intensive and potentially contributing to the climate problem which drives it. But preferably they will learn to reduce the consumption of, and recycle and reuse, their existing water resources.

An economically, environmentally and socially sustainable technology for water treatment should be cheap and energy-efficient, with little or no chemical consumption, it should facilitate water recycling and reuse that minimizes the direct disposal of wastewater to the aquatic environment, and it should be a technology which can be easily accessed and deployed over a wide range of physical scales. There are many existing technologies which compete to achieve these aims in specific contexts, such as electrodialysis, membrane filtration, adsorption and precipitation, electrochemical reduction and electro-deionization. However, the same technologies may also consume large amounts of energy and, in the process, transfer pollutants between the different fluid phases, wastes, and by-products which are generated.

Since the early 1970s, photocatalytic advanced oxidation processes via heterogeneous semiconductor materials have emerged as a viable technology for the objectives of sustainability, as well as overcoming the aforementioned limitations, and have been the subject of intensive research - in particular, in water/air purification and water splitting. In this process, photons with energy equal to or greater than the band gap of the material are adsorbed by the particulate catalyst, and this results in the formation of a negative conduction band electron and a positive valence band hole. Both of which can participate in a variety of redox reactions in water treatment, but hydroxyl radicals (both surface and bulk) are often considered the dominant oxidant.

The advantages of photocatalysis over other homogeneous-phase AOPs are well documented. When operated under mild conditions as a tertiary treatment, it offers a simple, low energy and sustainable technology which is able to degrade the persistent organic pollutants still remaining in wastewater following conventional biological and physical treatments into water, carbon dioxide and other small molecules. It can also reduce or oxidize inorganic pollutants to harmless substances, and inactivate microorganisms as an effective disinfection process. The benchmark commercial material, titanium dioxide, is cheap, non-toxic and robust. It requires low energy ultraviolet light for its excitation, promising solar applications. It also avoids the need for the supply of treatment chemicals, which is a strong advantage in remote or resource-limited applications. Recently, a surge in articles in the scientific literature – with over 10,000 research articles published in the last twenty years – underwrites this positive image.

Foreword

Nevertheless, considerable technology transfer problems remain, perpetuating a widening gap between academic vision and industrial application. This timely treatise therefore tries to throw a bridge across this gap. As with many innovations, a well-defined pathway for transferring this unique technology to industry will allow engineers to reap the potential.

An introduction to photocatalysis is provided in Chapter 1. Applications are outlined to water and wastewater treatment, as well as to photolysis for energy production. An overview of the mechanisms and characterization of charge generation are provided, and strategies for improving photocatalytic activity are discussed.

Chapter 2 presents metal organic frameworks as an emerging vehicle to overcome the limitations of established photocatalysts. Mechanisms of action and performance data are presented for a variety of pollutants in aqueous environments, and their rather unique features suggest novel composites and functionalized structures. Their essential features, superior efficiency and critical limits are reviewed.

In Chapter 3, the focus now switches to the photocatalytic reactors themselves, and their different types and configurations. The treatment process results from the interaction of three main components: pollutants, catalysts and source of photons. It is the differences in these components across a wide range of physical scales that make reactor design in photocatalysis so challenging and at the same time so promising.

Photocatalysis is now showing great promise for the treatment of various wastewater streams such as toxic landfill leachates, which is the subject of Chapter 4. Such leachates are extremely deleterious to the environment. After analyzing the characteristics of the leachates, homogeneous and heterogenous treatment processes employing photocatalysis are reviewed. An investigation is made of the effect of reactor operating parameters, and finally the performance of different types of reactor configuration is evaluated.

A life-cycle analysis (LCA) is essential for the development of any sustainable process for water treatment in practice, and such an analysis is presented in Chapter 5 for a solar-driven photocatalytic process for wastewater treatment. Following an overview of the LCA framework, the goal, the scope and the system boundary are all defined. Titanium dioxide as a ubiquitous photocatalyst is given special attention. Key findings on the environmental impacts for solar-driven photocatalytic wastewater treatment are presented.

In the final two chapters, an analysis of patents in the photocatalytic treatment of water and wastewater is presented. Such an analysis helps to reveal insights into the ongoing work in a particular field of technology.

In Chapter 6, photocatalytic materials are first presented. Following an overview of trends in patents, an analysis of patent registration over time and the activity of key players is made. This suggests that Japan first established itself as a pioneer, but nowadays China has assumed the dominant role. Solar energy and nanotechnology are seen as two areas in particular that have received increasing attention in the patent arena.

Finally, in Chapter 7, the patent focus thus switches to solar energy and nanotechnology. Registration and destination data regarding the key players are analyzed, and once again the dominance of China (and her universities in particular), Japan and the USA in this area is revealed. Today, titanium dioxide remains the most widely used nano-material for photocatalysis. In the patent landscape, it is seen that research institutions rather than industrial companies are leading the way.

When Dr Alireza Bazargan invited me to write this foreword, I was very happy to oblige. I remember an Iranian friend of mine once told me a beautiful story about the Iranian sense of honour. I had learned that in Persia, a transaction could be secured on the basis of a single human hair given by the lender to the creditor, which was returned to the lender on completion, without the need for a contract. This meant that a single strand of hair could be used as a form of honourable security. So with that, Ali, I graciously return your strand.

Professor Nicholas P. Hankins
MA PhD CEng FIChemE MRSC PGCAP, The Oxford Centre for
Sustainable Water Engineering, Department of Engineering Science,
The University of Oxford, United Kingdom.
Oxford, March 2022.

About the Editor

Dr. Alireza Bazargan was born in Tehran, Iran, and moved to Canada with his family at a young age. After spending most of his childhood and adolescence in Toronto and Vancouver, he returned to Iran and soon after participated in the highly competitive nationwide university entrance exam. After being ranked in the top percentiles among hundreds of thousands of participants, Dr. Bazargan was able to gain admission into the Chemical Engineering Department of Sharif University of Technology.

Near the end of his BSc studies, Dr. Bazargan was awarded a scholarship by IAESTE/DAAD to carry out a research internship at Technische Universität Kaiserslautern in Germany. Subsequently for his Masters, Dr. Bazargan was the recipient of the esteemed TOTAL scholarship allowing him to enter École Nationale Supérieure des Mines de Nantes (now IMT Atlantique) which is one of France's elite grande écoles, to obtain a degree in Project Management for Environmental and Energy Engineering (PM3E).

In order to complete his MSc thesis, Dr. Bazargan received a scholarship from the Hong Kong University of Science and Technology (HKUST), ranked as one of the top 40 universities in the world. Thereafter, Dr. Bazargan received admission from Oxford University for conducting his PhD under the supervision of Prof. Nick Hankins. As fate would have it, the trip to Oxford did not come to

© 2022 The Editors. This is an Open Access book chapter distributed under the terms of the Creative Commons Attribution Licence (CC BY-NC-ND 4.0), which permits copying and redistribution for noncommercial purposes with no derivatives, provided the original work is properly cited (https://creativecommons.org/licenses/by-nc-nd/4.0/). This does not affect the rights licensed or assigned from any third party in this book. The chapter is from the book Photocatalytic Water and Wastewater Treatment, Alireza Barzagan (Ed.).

pass, but it planted the seed from which the relationship between Dr. Bazargan and Prof. Hankins formed.

Next, Dr. Bazargan accepted a full scholarship from HKUST, in order to continue his research under the supervision of one of the world's most cited chemical engineers, Prof. Gordon McKay. As part of his doctorate, Dr. Bazargan received an additional scholarship to carry out a portion of his work at the University of Cambridge, UK. At Cambridge, Dr. Bazargan was a member of the Paste, Particle and Polymer Processing Group (P4G) in the Department of Chemical Engineering and Biotechnology. Dr. Bazargan received his PhD in January 2015.

Currently, Dr. Bazargan is a faculty member at the School of Environment, University of Tehran, where since 2019 he has been the Head of the Waste Management Research Group as well as a member of the Water and Wastewater Treatment Group. In addition to academic duties such as teaching and supervising graduate students, during his time at the University of Tehran, Dr. Bazargan has founded the Resource Efficiency Laboratory which he currently directs.

As a proponent of a lively academic-industrial interface, Dr. Bazargan co-founded Pyramoon Water and Energy Engineering Company in 2019, which strives to develop novel environmental technologies. The company's accomplishments in the span of just a few years since its inception include the construction of several desalination plants (cumulative capacity of over 20 million liters per day), and the design and development of numerous products such as Dissolved Air Flotation (DAF) with embedded filtration, novel vortex separation units for grit removal, and 3D-printed static mixers to name a few.

Throughout his career, Dr. Bazargan has been the recipient of numerous awards and honors including awards for best presentation, best paper, top instructor and a gold medal among numerous others. In 2015, Dr. Bazargan was inducted into Iran's National Elites Foundation. In the same year, he won the International Young Waste Researcher Award. In 2017, he was a recipient of the Kazemi Ashtiani prize, and in 2018, his first edited book *A Multidisciplinary Introduction to Desalination* became a #1 New Release Best Seller on Amazon. com. Since 2020, Dr. Bazargan has also been a contractor and consultant to the United Nations Development Program, whereby he provides the UN offices in Iran with expert opinion.

During the years, Dr. Bazargan has provided professional services to numerous companies and organizations including The Science and Technology Vice Presidency of the Government of Iran, The Nanotechnology Initiative Council of Iran, The Golrang Industrial Group, Hard Tech Startup Accelerator, Noor Vijeh Company, and Peako STEP Ltd as its exclusive representative in Iran for Biomass Gasification units. Dr. Bazargan is a native speaker of English and Farsi, has a working proficiency in French, and a basic understanding of Mandarin Chinese and Arabic. Naturally, his colorful past and experiences have made him uniquely suited for collaborations with international researchers and multinational companies alike.

doi: 10.2166/9781789061932_0001

Chapter 1
An introduction to photocatalysis

Mohammad Reza Boskabadi[1], Vincent Rogé[2], Alireza Bazargan[3], Hamed Sargazi[3] and Emanuele Barborini[2*]

[1]School of Chemical Engineering, College of Engineering, University of Tehran, Tehran, Iran
[2]The Nanomaterial and Nanotechnologies Unit, Materials Research and Technology Department, Luxembourg Institute of Science and Technology, Luxembourg
[3]School of Environment, College of Engineering, University of Tehran, Tehran, Iran
*Corresponding author: emanuele.barborini@list.lu

ABSTRACT
In this chapter the principles behind photocatalytic water and wastewater treatment as well as water splitting for energy production (producing oxygen and hydrogen gas) are discussed. The mechanisms involved in the generation of charge carriers in semiconductors and their behavior towards pollutant degradation or water splitting are also outlined. As conferred, a variety of organic molecules have been degraded by photocatalysis, including dyes, phenols, nitrogen-containing molecules, pesticides, and pharmaceuticals. Characterization techniques with which lifetimes and spatial distributions of charge carriers are displayed in photocatalytic materials are touched upon. Additionally, various strategies for improving the photocatalytic activity are discussed as well as the reasons behind the development of such strategies. This includes describing the formation of heterostructures formed between semiconductors as well as the experimental parameters that affect the kinetics of such reactions. A quick summary of the operational parameters that affect the photocatalytic process reviews the effects of pH, temperature, presence of oxidants, concentration of the target pollutant, catalyst loading, and light intensity and wavelength.

1.1 PRINCIPLES OF PHOTOCATALYSIS

With increasing worldwide interest in environmental issues such as global warming, water/air pollution and waste management, many efforts are being made to find cost effective and sustainable processes for energy production, pollution elimination or recycling. The most abundant energy source available on earth is sunlight. Consequently, engineering efficient solar technologies is

© 2022 The Editors. This is an Open Access book chapter distributed under the terms of the Creative Commons Attribution Licence (CC BY-NC-ND 4.0), which permits copying and redistribution for noncommercial purposes with no derivatives, provided the original work is properly cited (https://creativecommons.org/licenses/by-nc-nd/4.0/). This does not affect the rights licensed or assigned from any third party in this book. The chapter is from the book Photocatalytic Water and Wastewater Treatment, Alireza Barzagan (Ed.).

a critical step towards carbon-free energy production. In principle, the solar energy received on earth every day should be enough to meet all of mankind's energy needs. However, solar systems are not yet efficient enough to convert and store the required energy. For this reason, solar energy is an exciting and rapidly growing research topic attracting scientists' interest in different fields, including electricity production with photovoltaic or photothermal solar panels, hydrogen production with photocatalytic/photo-electrocatalytic water splitting, and solar photocatalytic water treatment.

Photocatalytic materials are materials that are able to convert an incident photon into a consumable or storable energy source, through the creation of an electron/hole pair at the photocatalyst level. From a general point of view, a photocatalyst is a catalyst that exhibits its catalytic properties under light irradiation, via absorption of photons. Such photocatalytic systems exist in nature, like in the photosynthesis process occurring in green leaves, where chlorophyll acts as a photocatalyst allowing the conversion of water and CO_2 into O_2 and sugars. Engineered photocatalysts tend to mimic this natural process by employing photogenerated electron/holes to create energetic radicals, which can be used for various applications such as water treatment or splitting water into O_2 and H_2.

The photocatalytic water treatment process is part of a broader family called advanced oxidation processes (AOPs). They all involve the production and utilization of highly oxidative OH· radicals to fragment and decompose organic molecules. Targeted molecules could be residual xenobiotics, dyes, pesticides or herbicides present in the wastewater due to human consumption or industrial activity [1]. AOPs are currently applied in many wastewater treatment plants, particularly photolytic and photo-Fenton processes. In photolytic processes, sacrificial molecules such as hydrogen peroxide, H_2O_2, or ozone, O_3, are irradiated under powerful germicidal ultraviolet (UV) light (254 nm) in order to produce oxidative radicals. In the photo-Fenton process, catalytic ferrous ions (Fe^{2+} or Fe^{3+}) are added with H_2O_2 under UV light, in order to generate OH· radicals more efficiently [2]. The drawback of these processes is that they use expensive and energy-consuming UV lights, and catalysts like ferrous ions operate suitably in a narrow acidic pH range (2.8–3.5) in order to avoid the precipitation of inactive iron oxyhydroxide species [3]. Thus, a pH neutralization of the treated water with alkaline chemicals is necessary.

To overcome those limitations, heterogeneous photocatalysis appears as a promising alternative. In this process, photocatalysts are solid materials like particles/nanoparticles suspended in water or supported on a substrate like a membrane. No sacrificial or additional chemicals are needed. The advantage of supported photocatalysts over the suspended ones is that no filtration process is needed afterwards. Nevertheless, the specific surface area of suspended photocatalysts is generally higher than that of supported ones, leading to improved degradation properties. Currently, the most effective photocatalysts operate in the UV range of the light spectrum. Thus, they suffer the same limitations as the photolytic or photo-Fenton processes cited above, that is the need for energy-consuming UV lights. UV light represents only a small percentage (5%) of the solar light passing through the atmosphere. Most of the

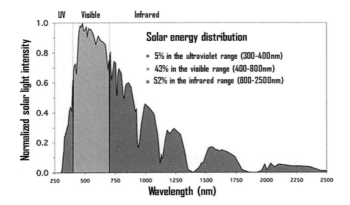

Figure 1.1 Spectrum representing the solar light received at the surface of the earth. Reprinted from [4].

solar light received at the surface of earth is in the visible (43%) and infrared region (52%) (Figure 1.1). Considering this, one can easily understand the crucial need for investing in research for the fabrication of visible-light active materials.

In addition to the ability to operate in the visible range, other parameters are being studied to enhance the performance of photocatalysts, like improving the lifetime of charge carriers and their mobility/availability at the surface.

1.2 MECHANISM AND CHARACTERIZATION OF THE PHOTOCATALYTIC PROCESS

1.2.1 Photocatalytic mechanism

The photocatalytic degradation process is schematically shown in Figure 1.2 [5]. Photons having energy larger than the photocatalyst's band gap are absorbed by the photocatalyst: an electron from its valence band is ejected into its conduction band, thus an electron/hole (e^-/h^+) pair is created. If the positive charge carrier

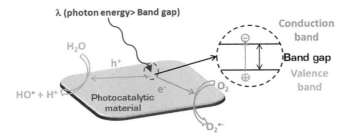

Figure 1.2 Schematic representation of a photocatalytic initiation process. Reprinted with permission from [5].

Photocatalytic Water and Wastewater Treatment

h^+ remaining in the valence band of the photocatalyst has an oxidation potential above 2.31 V/NHE (normal hydrogen electrode), (potential of the couple OH·/ H^+, H_2O) at pH$=0$, it can oxidize water into H^+ and OH·. If the negative charge carrier e^- remaining in the conduction band of the photocatalyst has a potential below 0.92 V/ENH (potential of the oxygen reduction in O_2^-) at pH$=0$ [6], it can reduce O_2 into O_2^-. With an oxidation potential of 2.31 V/ENH at pH$=0$, OH· radicals are energetic enough to break covalent C−C bonds in organic molecules, while O_2^- radicals have a reductive potential sufficient to reduce water into hydrogen peroxide (H_2O_2). Hydrogen peroxide decomposes under UV light into 2 OH·. The photo-generation of OH· is direct during the water oxidation with h^+, while it is a two-step process during the water reduction with e^-. Consequently, photocatalytic degradations of organic compounds are more efficiently driven by the oxidative OH· generation with h^+.

The overall mechanism for the photocatalytic degradation of organic compounds can be written as follows:

Photocatalyst$+$hv$=$Photocatalyst $(h_{(vb)}^+ + e_{(cb)}^-)$

$$H_2O + h_{(vb)}^+ \rightarrow H^+ + OH· \tag{1.1}$$

$$e_{(cb)}^- + O_2 \rightarrow O_2^- \tag{1.2}$$

$$O_2^- + H^+ \rightarrow HO_2· \tag{1.3}$$

$$O_2^- + HO_2· + H^+ \rightarrow H_2O_2 + O_2 \tag{1.4}$$

$$H_2O_2 + h\nu \rightarrow 2\,OH· \tag{1.5}$$

$$OH· + \text{organic compounds} \rightarrow CO_2 + H_2O + \cdots \tag{1.6}$$

The photocatalytic water splitting follows the same mechanism. In this case, photocatalysts must generate negative charge carriers e^- with oxidation potentials more reductive than 0 V/NHE at pH$=0$ (the oxidation potential of the couple H^+/H_2); and positive charge carriers h^+ with oxidation potentials more oxidative than 1.23 V/NHE at pH$=0$ (the oxidation potential of the couple O_2/H_2O). The overall water splitting can be described as follows [7]:

$$2H_2O \rightarrow 2H_2 + O_2(\Delta G° = 238\,kJ.mol^{-1}) \tag{1.7}$$

With half-reactions:

$$2H^+ + 2e_{(cb)}^- \rightarrow H_2 \tag{1.8}$$

$$2H_2O + 4h_{(vb)}^+ \rightarrow O_2 + 4H^+ \tag{1.9}$$

With a large positive change in the Gibbs free energy (238 kJ.mol⁻¹), water splitting is an uphill reaction that needs to be activated with catalysts (or photocatalysts). Since the energy needed to split the water molecule is 1.23 eV (1.23 V/NHE at pH$=0$), the water splitting process can theoretically be performed with photocatalysts absorbing photons in the infrared range

An introduction to photocatalysis

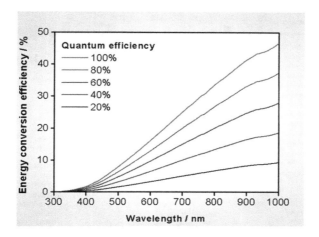

Figure 1.3 Calculated energy conversion efficiency versus wavelength for different quantum efficiencies. Reprinted from [8].

(indeed photon energies higher than 1.23 eV correspond to wavelengths smaller than 1010 nm). Based on calculations using fixed quantum efficiencies of photocatalysts, K. Maeda [8] showed that the solar energy conversion effectiveness increases with the increasing wavelength of photons, from the UV to the infrared part of the light spectrum (Figure 1.3). This is explained by the highest number of photons available when the wavelength increases (Figure 1.1). Results presented in Figure 1.3 endorse the necessity of developing visible light active photocatalysts.

1.2.2 Lifetime and mobility characterization of photo-excited charge carriers

As the photocatalytic process requires the photo-production of e^-/h^+ pairs and their interaction with the target molecules, the understanding of pair recombination times and mobilities is of paramount importance in order to rationally design efficient photocatalysts. The global timescale of a photocatalytic process extends over hundreds of milliseconds to a few seconds, as presented in Figure 1.4 [9]. Typically, photon absorption and the formation/diffusion of charge carriers e^-/h^+ are very quick steps performed within a few picoseconds (ps). Then, the charge carriers diffuse within nanoseconds (ns) to microseconds (µs). In several materials, the nanosecond to microsecond range is also the timescale range for recombination of charge carriers [10, 11]. The oxidation/reduction process, following the transport of charge carriers at the surface of the photocatalyst and involving the mass transfer of the water/pollutants at the interface, is the last process occurring in a few hundreds of milliseconds (ms) to seconds (s).

In the photocatalytic process mechanism, lifetimes and mobilities of carriers depend on multiple factors like the chemical environment, the driving force of transport energetic charge to absorbed reactant molecules, and the

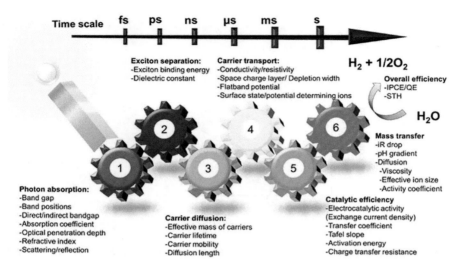

Figure 1.4 Timescale for photocatalytic water splitting. IPCE: incident photon to current efficiency, QE: quantum efficiency; STH: solar-to-hydrogen. Reprinted from [9].

lifetime of highly reactive intermediates [11]. Experimental techniques able to precisely characterize excitation/recombination kinetics (ps-μs), charge separation and transfer (ps-ns), surface reactions (μs-ms) or diffusion mean free path have emerged in recent years. Among them, one can distinguish optical spectroscopies and electron or scanning probe microscopies, like time resolved photoluminescence/fluorescence (tr-PL), transient absorption spectroscopy (TAS), time resolved microwave conductivity (TRMC), ultra-fast electron microscopy, kelvin probe force microscopy (KPFM) or conductive atomic force microscopy (C-AFM).

Optical processes rely on the principle of time resolved spectroscopy, like pump-probe transient absorption spectroscopy, time resolved microwave conductivity or time resolved photoluminescence. In the transient absorption spectroscopy process, an incident energy source (usually a laser pulse in the visible or infrared range) acts as a pump that excites atoms or electrons and generates a non-equilibrium state. A second light source, weaker than the first one and delayed in time, probes the induced changes in optical constants. It allows the determination of the relaxation time of generated species, with a time resolution directly linked to the pump source used [12]. Thanks to the latest laser technologies, able to deliver laser pulses below 10 fs, the time resolution offered by ultra-fast transient absorption spectroscopy allows the characterization of the early stages of the photocatalytic processes (fs to ns).

Time resolved microwave conductivity is a variant of transient absorption techniques (involving visible or infrared light sources). In this case, microwave or terahertz electromagnetic waves are used as pump sources. The major difference

An introduction to photocatalysis

compared to visible/infrared excitation is that the absorption is directly related to the conductivity of charge carriers created in the material [13], and not to optical transitions from the ground state to non-equilibrium ones. This allows the qualitative and quantitative measurement of changes in microwave absorption resulting from light induced production and decay of charged and dipolar molecular entities [14]. The microwave absorption intensity is a function of the product of the number of mobile charges and their mobilities. This characterization technique unveils evidence on recombination of charge carriers and dynamics, effective mobility, and lifetime and trapping of formed carriers [15], within the femtosecond to nanosecond scale (depending on the probe laser).

Time resolved photoluminescence is a technique close and often complementary to transient absorption spectroscopy. It measures the emitted fluorescence or photoluminescence signal instead of changes in absorption properties (for TAS). Just like transient absorption, the excitation source is a short laser pulse creating non-equilibrium states in the materials [11]. The radiative decay of charge carriers, due to recombination or trapping, is detected with a time resolution from a few picoseconds to microseconds [16, 17]. The complexity of the full photocatalytic process, due to the large timescale involved between the light absorption and the modification of the target molecules (e.g. water splitting or degradation of organic compounds), makes its complete characterization a difficult task. As an example, Dillon *et al.* revealed that four different samples of TiO_2@Au core/shell structures, which were showing different efficiencies toward water splitting, had similar TAS kinetic curves with a longer time delay of around 1.5 ns [18]. This result can be explained by the fact that, at the timescale studied (the nanosecond scale), all four samples had the same properties, but most likely photo-excited charge carriers with longer lifetimes were responsible for the water splitting. Consequently, the study of transient absorption at the nanosecond timescale alone can account for only a part of the photocatalytic process. In order to understand the photocatalytic properties of carbon nitride, R. Godin *et al.* [19] investigated charge trapping and recombination with tr-PL and TAS techniques, over 10 orders of time magnitude. This allowed them to elucidate fundamental excited state processes that dictate the global photocatalytic activity.

In addition to characterization of the carriers' lifetimes and mobilities, the investigation of their spatial distribution and transfer is key to precisely determine active sites on photocatalysts. To this end, AFM-based techniques, such as KPFM or C-AFM, have been shown to be powerful tools. KPFM measures electrostatic forces between the sample and a cantilever, leading to the description of the electric potential distribution of sample surface with nanometric lateral resolution. It is usually adopted in materials science for work function mapping [20] or dopant profile determination within semiconductors [21]. In addition, it can also unveil surface photovoltage under illumination (difference in contact potential under illumination and in the dark) and, consequently, achieve spatially resolved surface photovoltage spectroscopy (SR-SPS) [22]. Using KPFM coupled with standard AFM, A. Jian *et al.* studied the electron transfer in an Au/TiO_2 nanoparticles/thin film system under light irradiation [23]. As depicted in Figure

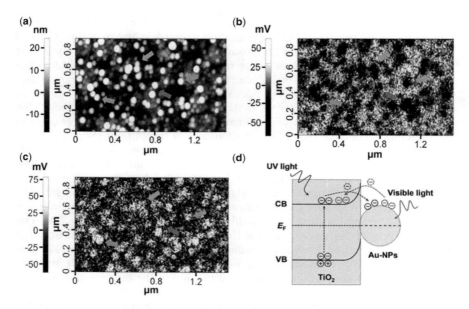

Figure 1.5 (a) AFM topography image, (b) surface potential images of Au/TiO$_2$ under UV light, (c) surface potential images of Au/TiO$_2$ under visible light. Arrows with same color identify the same particles in these three images, (d) scheme of the electron migration mechanism under UV or visible light. Figure reproduced from [23].

1.5, the surface charge of gold particles under visible or UV light is determined in situ by KPFM. This technique revealed that under UV light, surfaces of gold nanoparticles were charged negatively (Figure 1.5b), whereas under visible light they were charged positively (Figure 1.5c). Consequently, KPFM measurements help to determine the electron migration mechanism in the Au/TiO$_2$ system (Figure 1.5d). Under UV light, electrons/holes are photogenerated in the TiO$_2$ film, and electrons in the conduction band migrate at the gold nanoparticle's surface. When the material is irradiated by visible light, the incident energy is too low to produce electron/hole pairs in the TiO$_2$, but a localized surface plasmons resonance (LSPR) phenomenon appears at the surface of the gold nanoparticles. Excited electrons from the LSPR migrate from the gold nanoparticle to the TiO$_2$ film, leading to a positive net charge on the gold surface.

C-AFM measures the local direct current (DC) between the sample and the cantilever. When operated in tapping mode (cantilever resonance oscillation), it offers the possibility of measuring the current–voltage behavior of the sample with nanometric lateral resolution [24], leading to large multidimensional datasets. C-AFM allows correlation between the morphology and the (photo) current response of the photocatalyst [25]. In particular, C-AFM enables the mapping of surface conductivities and charge transfer characteristics at grain boundaries, facet planes or at the interface between two different materials (e.g. heterojunction, co-catalyst). Recently, Eichhorn *et al.* [26] showed that C-AFM

An introduction to photocatalysis

measurements, using back side illumination, allowed for the determination of optoelectronic heterogeneities in $BiVO_4$ photoanodes.

1.2.3 Photocatalysis characterization by organic dye degradation

When one thinks about photocatalytic water treatment, organic dyes, pesticides, herbicides and xenobiotics appear among the most important molecules to eliminate. If the spatiotemporal characterization of charge carriers at the nanometric-microscopic level provide useful pieces of information on photocatalyst properties, macroscopic processes are necessary in order to precisely determine their actual photocatalytic performances. Organic dyes are among the most studied molecules for the determination of photocatalyst degradation performances. Textile and manufacturing industries are responsible for water pollution with many different types of colored dyes, and their elimination is a serious problem to tackle. Dyes can be classified based on their chromophoric group, including azo dyes, acridine dyes, nitro dyes, phenothyazine dyes, xanthene dyes or quinine-amine dyes [27]. Thanks to their chromophoric groups, the photocatalytic degradation of organic dyes can easily be tracked with visible light absorption measurements. The degradation of dyes on photocatalysts usually follows a pseudo-first order reaction kinetic: the Langmuir-Hinshelwood model [28]. This model can be expressed as follows:

$$r = \frac{dC}{dt} = k_r q_x = \frac{k_r KC}{1 + KC} \tag{1.10}$$

in which r denotes the photodegradation rate. The surface on the photocatalyst covered by the target molecules is shown with q_x. The parameters k_r and C are the reaction constant and concentration at time t. K is the adsorption constant which depends on the surface of the catalyst as well as the molecule. If the equation is rearranged it will look as follows:

$$dt = \frac{1 + KC}{k_r KC} dC \tag{1.11}$$

Or,

$$t = \frac{1}{k_r K} \ln \frac{C_0}{C} + \frac{1}{k_r} (C_0 - C) \tag{1.12}$$

Under conditions where the initial concentration of C_0 is negligibly low, Equation (1.12) can be simplified to the following form:

$$\ln \frac{C_0}{C} = k_r Kt = k't \tag{1.13}$$

where k' corresponds to the pseudo-first order constant reaction. In order to simplify the reaction kinetics to the first order, most of the photocatalytic degradation processes of dyes involve low initial concentration and thus the photocatalytic degradation constant is determined based on Equation (1.13).

Photocatalytic Water and Wastewater Treatment

Rhodamine B (xhantene) [29, 30], methylene blue (phenothiazine) [5, 31] and methyl orange (azo dye) [32] are common dyes used to probe photocatalytic reactions. Their total mineralization (full degradation of the molecule to H_2O and CO_2) can be obtained via the photocatalytic degradation process on photocatalysts [33, 34]. However, not all dyes can be easily mineralized. As an example, dyes containing the triazine group tend to form highly stable cyanuric acid [35]. During the photocatalytic degradation process with colored dyes, the disappearance of the color is usually attributed to the degradation of the molecule. This consideration is often erroneous, as the decoloration is just the consequence of the degradation of the chromophoric group, and not of the full molecule. In azo dyes, the decoloration appears after the attack of the azo bond $(C-N=N-)$ [36]. This is followed by aromatic rings opening, then formation of carboxylic acids which ultimately decarboxylate by the 'photo-Kolbe' reaction to release CO_2 [37]. Consequently, the decoloration of dyes can provide indications of photocatalyst performances, but does not signify the complete mineralization of the molecule. In order to improve the process characterization, total organic carbon (TOC) analysis, which quantitatively determines the amount of carbon in organic molecules present in the water, is often used [38, 39].

1.3 HOMOGENEOUS AND HETEROGENEOUS PHOTOCATALYTIC PROCESSES

Generally, homogeneous systems are those in which the catalyst is in the same phase (gas or liquid) as the reactants. Conversely, in heterogeneous systems, the catalyst is not in the same phase as the reactants, such as the use of solid and/or powdered catalysts in a liquid reaction mixture. For example, the heterogeneous reaction employing TiO_2/UV and the homogeneous reaction using Fe^{2+}/H_2O_2 have attracted a lot of interest [40].

1.3.1 Homogeneous example: photo-Fenton reaction

Fenton's reaction is named after Henry Fenton, who observed the activation of H_2O_2 in the presence of iron for the oxidation of tartaric acid [41]. The classical mechanism of the photo-Fenton process is a simple redox reaction in which Fe^{2+} is converted to Fe^{3+} and H_2O_2 to a hydroxyl ion (OH^-) and hydroxyl radical $(OH\cdot)$ with a strong oxidation potential $(E_0 = 2.73$ V vs. NHE) [42]. Following this reaction (Equation (1.14)), the generated ferric ion can be converted back to ferrous ions by H_2O_2 (Equation (1.15)). The reduction of Fe^{3+} as per Equation (1.15) with a reaction constant of 0.02 $M^{-1}s^{-1}$ is much slower than the oxidation of Fe^{2+} as per Equation (1.14) with a reaction constant of 40–80 $M^{-1}s^{-1}$. Thus, the rate limiting step is identified, and large amounts of initial Fe^{2+} ions are required to mineralize organic pollutants [43].

$$Fe^{2+} + H_2O_2 \rightarrow Fe^{3+} + OH^- + OH\cdot \tag{1.14}$$

$$Fe^{3+} + H_2O_2 + H_2O \rightarrow Fe^{2+} + H_3O^+ + HO_2^- \tag{1.15}$$

An introduction to photocatalysis

$$OH\cdot + H_2O_2 \rightarrow HO_2^{\cdot -} + H_2O \tag{1.16}$$

$$OH\cdot + Fe^{2+} \rightarrow Fe^{3+} + OH^- \tag{1.17}$$

Many studies have focused on the photo-Fenton treatment of water and wastewater. For example, Ahile *et al.* (2021) sought to answer the question 'Is iron-chelating in a homogeneous photo-Fenton process at neutral pH suitable for purifying the output current from the biological secondary purification system?'. In the study, oxalic acid (OA), nitrileutria acetic acid (NTA), ethylenediamine acid (EDDS), citric acid (CA), and ethylenediaminetetraacetic acid (EDTA) were used as chelating agents at pH = 7 while a homogeneous photo-Fenton process took place. The results were reported based on disinfection and bacterial regrowth leading to the preferential order of EDTA > OA > NTA > CA > EDDS. It was also observed that all iron chelates caused an increase in chemical oxygen demand (COD) in the effluent, for which the effect of EDDS was higher than other compounds [44].

1.3.2 Heterogeneous photocatalytic reaction examples

With the discovery of the photocatalytic properties of TiO_2, research regarding heterogeneous photocatalysis has become commonplace. The heterogeneous photocatalytic mechanism begins primarily with the ability of semiconductors to produce charge carriers. This is followed by the production of hydroxyl free radicals, which eventually leads to the breakdown of organic compounds to H_2O and CO_2 [45]. The most attractive features of this process are [46]:

- Pollutants can be completely degraded to CO_2 and other minerals
- The process usually takes place at ambient conditions
- The only requirement for the reaction to start is the presence of oxygen and energy received above the band gap energy, both of which are abundantly available
- Various types of inert matrices, including glass, polymers, carbon nanotubes, and graphene oxides, can be used as catalyst supports
- The catalyst is inexpensive, non-toxic, and reusable

The heterogeneous photocatalytic mechanism involves a chain of oxidation and reduction reactions at the photocatalyst level. In a semiconductor, the distance between the last electron-occupied band (capacitance band) and the first empty electron band (conduction band) is called the band gap. When the energy of the photons colliding with the semiconductor is greater than or equal to the energy of the gap band, the electrons in the capacitance band are excited and migrate to the conduction band within a few femtoseconds. The electron void in the valence band is itself an electrical conductor, so this process leads to electron charge carriers and holes. If these electrons and holes are trapped on the surface of the semiconductor and the recombination of the electron/hole pair is prevented, the following set of reactions will occur:

1- Photon excitation	$TiO_2 + hv \rightarrow e^- + h^+$
2- Trapping of free electrons	$e^-_{CB} \rightarrow e^-_{TR}$
3- Stuck holes	$h^+_{VB} \rightarrow h^+_{TR}$
4- Reassembly of charge carriers	$e^-_{TR} + h^+_{VB} \rightarrow e^-_{CB} + heat$
5- Sweeping the excited electrons	$(O_2)_{ads} + e^- \rightarrow O^-_2$
6- Hydroxyl oxidation	$OH^- + h^+ \rightarrow OH\cdot$
7- Photocatalytic degradation with hydroxyl radical	$R\text{-}H + OH\cdot \rightarrow R' + H_2O$

The hydroxyl radicals produced in step 6 convert organic impurities into intermediates, which are often converted to water and carbon dioxide as a by-product by the same reaction or reactions [45].

Biancullo *et al.* (2019) investigated the effect of a heterogeneous TiO_2 photocatalyst for the treatment of effluent from a secondary treatment system with azithromycin (AZT), trimethoprim (TMP), ofloxacin (OFL), and sulfamethoxazole (SMX). In this study, various operational parameters such as irradiation conditions, amount of catalyst, and use of methanol as a carrier solvent and radical scavenger were studied. The most efficient conditions for municipal wastewater treatment (four light emitting diodes (LEDs) symmetrically distributed and a catalyst concentration of $1\ g.L^{-1}$) were used to remove antibiotics in real conditions. The results showed that one hour of photocatalytic treatment was sufficient to reduce antibiotics to less than the allowable level [47].

Elsewhere, Ayekeo *et al.* (2019) studied the combined effect of coagulation-flocculation processes and heterogeneous photocatalytic activity to improve the removal of humic substances in water treatment from a river in Africa. In this study, heterogeneous photocatalytic activity was performed with the help of TiO_2-P25 suspended catalyst and TiO_2-P25/-SiC support material. Coagulation-flocculation and the heterogeneous photocatalytic method were used separately and together to evaluate the best removal process for organic compounds. The initial concentration of dissolved carbon (DOC) in the river water was about $20\ mg.L^{-1}$. To remove the compounds, the coagulation-flocculation process was first optimized. Optimization of the coagulation-flocculation process was achieved at pH = 5 with a dose of $110\ mg.L^{-1}$ of coagulant, which resulted in the removal of 70% of the humic substances. Coupling of the coagulation process and 220 minutes of photocatalytic irradiation resulted in the removal of an additional 80% of the humic substances which had remained after coagulation-flocculation. Therefore, with this combined process of coagulation-flocculation and heterogeneous photocatalytic activity, approximately 90% of the humic substances were removed [48].

1.4 APPLICATIONS OF PHOTOCATALYTIC PROCESSES

1.4.1 Water splitting

The global thirst for energy is one of the problems facing current and future generations. Among the various energy substitution strategies for fossil fuels, hydrogen is one of the main candidates; but the main problem is the lack of access to hydrogen gas in nature. Therefore, technology is needed to decompose and/or extract hydrogen from various materials which contain it, in a safe and clean manner. To this end, solar energy can be used to produce environmentally friendly hydrogen through water splitting. To be more specific, photocatalytic or photoelectrochemical water splitting techniques can be used to produce hydrogen using solar energy [49, 50].

For instance, Shi *et al.* (2019) investigated the photocatalytic effect of $Co_3(PO_4)_2/g\text{-}C_3N_4$ for hydrogen production by direct deposition of water under the effect of electrostatic colonic interaction. The presence of $Co_3(PO_4)_2$ plates increased the light absorption range of the unconventional $Co_3(PO_4)_2/g\text{-}C_3N_4$ heterostructure, and increased the contact surface which led to enhanced interfacial charge transfer between $g\text{-}C_3N_4$ and $Co_3(PO_4)_2$ nanoplates. The optimal heterostructure showed a maximum production rate of 375.6 and 177.4 $\mu mol.g^{-1}.h^{-1}$ for O_2 and H_2, respectively. In addition, the composite has high stability and high recyclability, which helps with the potential application of this structure for sustainable energy production [51].

1.4.2 Solar energy

The sun, as a natural nuclear reactor, releases small packets of energy, that is photons. These particles contain an enormous amount of energy that is sufficient for a large part of the earth's energy needs. Photocatalytic materials are used in various forms to absorb photon energy and generate electricity in solar cells. Many types of solar cells, such as organic, photoelectrochemical, color-sensitive, and hybrid cells, have been developed to use solar energy.

1.4.3 Reduction of carbon dioxide

It is no secret that burning fossil fuels leads to the production of CO_2, which is a greenhouse gas that slows down the escape of heat from the atmosphere into space and warms the earth. Many efforts are underway to reduce the amount of carbon dioxide in the atmosphere, including the absorption, storage and use of carbon dioxide. The conversion of CO_2 into value-added chemicals is one of the most attractive ways to reduce carbon dioxide in nature. However, the main problem with carbon dioxide conversion is its very stable structure and the high energy of the $C = O$ bond (805 KJ.mol^{-1} bond enthalpy). The use of photocatalysts to solve this problem has attracted the attention of many researchers [52]. The process mainly consists of three stages: light absorption, separation and transfer of charge carriers, and reduction reactions [53].

Wang *et al.* (2020) produced $FeCoS_2\text{-}CoS_2$ bilayer nanotubes to reduce carbon dioxide using the photocatalytic method under visible light. In the production strategy, two cation exchange steps connected two metal sulfides to a two-shell cylindrical heterostructure, with each shell being in the form of a two-dimensional nanoplate. By using the $FeCoS_2$-CoS_2 hybrid structure, the energy required to excite the charge carriers was reduced, facilitating their separation. In addition, this composite structure increased the active photocatalytic sites to reduce carbon dioxide and increase the light absorption efficiency. As a result, this double-walled nanotube showed high activity and stability [54].

1.4.4 Water and wastewater treatment

Various types of pollutants have contaminated water resources in recent decades. Emerging organic pollutants and resistant organic and inorganic compounds are among the pollutants that conventional water and wastewater treatment processes have a hard time removing with suitable efficiency [55]. New, and sometimes costly methods, are being explored to remove these pollutants, highlighting the need for economical and effective solutions. Thus far, the application of photocatalytic methods to remove pigments, phenols, nitrogen compounds, sulfur compounds, pharmaceutical compounds, pesticides and many other compounds has been investigated [56].

1.4.4.1 Dyes

In various dyeing processes, between 1% and 20% of the total production dye is lost and enters the environment as an effluent [57, 58]. Due to the nature of dyes, conventional biological treatment methods are not very effective [59]. Under anaerobic conditions, dyes are readily converted to amino aromatic compounds [60]. The method of adsorption and coagulation will also lead to the formation of secondary pollutants. However, methods based on photocatalytic processes have shown promising results for rapid and non-selective oxidation of organic dyes [61].

Kansal *et al.* (2009) investigated the degradation of reactive black 5 (RB5) and reactive orange 4 (RO4) dyes using a heterogeneous photocatalytic process. In this study, the photocatalytic activity of various semiconductors such as TiO_2 and ZnO was investigated. The experiments were performed by changing the pH range (3–11), the amount of catalyst (0.25–1.5 g.L^{-1}), and the initial dye concentration. The performance of the ZnO/UV photocatalytic system was better compared to the TiO_2/UV system. Complete decolorization of RB5 with ZnO occurred after 7 min, while only 75% decomposition occurred with TiO_2 after 7 min. For RO4 at the same duration, 92% and 62% decolorization were reported, respectively [62].

1.4.4.2 Phenols

Removal of phenolic contaminants by conventional methods such as adsorption with the help of activated carbon, membrane filtration, ion exchange, and so on. leads to the production of a condensed wastewater stream during the treatment process, and as a result, secondary treatment steps will be required [63]. These additional steps incur additional costs and environmental hazards. In recent years, the process of photocatalytic oxidation to remove these pollutants has yielded promising results [64].

For example, the photocatalytic degradation of resorcinol in a ZnO batch reactor has been investigated under visible light [65]. In this study, the effect of pH on COD reduction was shown to be significant, with neutral or alkaline pH proving more suitable. Mediators of the photocatalytic reaction were identified by Fourier transform infrared spectroscopy (FTIR) and gas chromatography coupled to mass spectrometry (GC/MS), as 1,2,4-trihydroxy-benzene and 1,2,3-trihydroxy-benzene. The final results showed complete removal of resorcinol (with an initial concentration of 100 ppm) and its mineralization on the surface of the ZnO photocatalyst by sunlight [65].

Elsewhere Lam *et al.* (2013) investigated the effect of TiO_2 and ZnO suspensions to evaluate the decomposition of phenol (ReOH) with an initial concentration of 10 mg.L^{-1}. The results showed that under optimal conditions, the removal efficiency of the contaminant by ZnO nanoparticles was better than by TiO_2. In the mechanism of degradation of ReOH phenol by TiO_2, the participation of hydroxyl radicals and positive cavities as oxidizing species were reported, while degradation in the presence of ZnO mainly occurred due to hydroxyl radicals [66].

In another study, $TiO_2/g\text{-}C_3N_4$ (TCN) thin-film electrodes were used by Wei *et al.* (2017) to investigate the combined effects of electrocatalysis and photocatalysis on the oxidation of phenolic contaminants with an initial concentration of 5 mg.L^{-1}. The study showed that phenol was completely oxidized in the presence of TCN with 1.5 V bias for 1.5 h under simulated sunlight [67].

1.4.4.3 *Nitrogen-containing compounds*
Nitrogenous compounds can cause environmental problems due to their high stability and solubility in water, as well as the fact that they are nutrients which can cause eutrophication.

Wang *et al.* (2010) investigated a photocatalytic treatment method for the treatment of nitrobenzene in industrial wastewater. In this study, a new photocatalyst consisting of a layer of phosphotogenic acid ($H_3PW_{12}O_{40}$) coated on a titanium dioxide (TiO_2) substrate was used. The results showed that the reaction time, the amount of catalyst, and the concentration of nitrobenzene were the main parameters determining the degradation of the pollutant. The maximum degradation of nitrobenzene with an initial concentration of 40 mg.L^{-1} in water was about 94.1% [68].

Elsewhere, Luo *et al.* (2015) synthesized and investigated the photocatalytic effect of $La/Fe/TiO_2$ on wastewater treatment containing ammonia nitrogen. In this study, $La/Fe/TiO_2$ catalyst was first synthesized using the sol-gel method. The produced composite had better chemical and physical properties in photocatalytic activity than pure TiO_2. Strong visible light response, higher contact surface, and more regular shape in morphology are some of the enhanced properties of this photocatalyst. The results of optical decomposition of ammonia nitrogen show that $La/Fe/TiO_2$ has higher catalytic activity for the degradation of ammonia nitrogen compared to pure TiO_2 and TiO_2 doped with a single metal. Ultimately, less than 70% of the ammonia nitrogen with an initial concentration of 100 mg.L^{-1} could be removed using the developed catalyst [69].

1.4.4.4 Sulfur-containing compounds

Sulfur compounds are toxic, corrosive and odorous, which can lead to environmental hazards. In various industries, including the oil and petrochemical industries, wastewater streams containing sulfur compounds are obtained from the desulfurization process. The presence of sulfide in wastewater reduces water-soluble oxygen and consequently endangers the life of living organisms in water [70]. Also, refineries need large amounts of hydrogen to refine sulfur-rich crude oils. To solve both problems, photocatalytic methods with high oxidation potential (+2.8 V) have been proposed to mineralize organic pollutants [71].

For example, Bharatvaj *et al.* (2018) doped Cerium (Ce^{3+}) onto titanium dioxide powder with the sol-gel method in order to reduce the TiO_2 band gap from about 3.2 to 2.7 eV (visible light region). The photocatalytic activity of this compound was evaluated by treating sulfidic wastewater [72].

1.4.4.5 Pharmaceutical compounds

Drug compounds are commonly found in municipal wastewater in concentrations from nanograms to micrograms per liter [73]. Although drugs are resistant to treatment, their low levels in municipal water and wastewater will not pose an operational problem for wastewater treatment plants. However, pharmaceutical wastewater is a distinct case that contains high levels of TOC. Photocatalytic methods are among the methods that are usually recommended for the treatment of these pollutants [74].

Deng *et al.* (2018) investigated the deposition of silver nanoparticles on semiconductors to improve the separation of electron-pores produced by light due to plasma resonance. The performance efficiencies of existing Ag-based photocatalysts are still low and practical applications are yet to materialize. In this study, silver nanoparticles were coated on $AgIn_5S_8$, using a solvothermal method and photon reduction, to improve the separation of carriers and the catalytic activity in the visible light. The 2.5% $Ag/AgIn_5S_8$ nanocomposite showed the highest photocatalytic degradation efficiency with 95.3% degradation efficiency for tetracycline hydrochloride with an initial concentration of 10 $mg.L^{-1}$. The catalyst was also used for real wastewater treatment in the pharmaceutical industry and showed an acceptable rate of mineralization and COD removal [75].

Degradation of carbamazepine (CBZ) and ibuprofen (IBP) in an aqueous environment with ZnO and TiO_2 photocatalysts under ultraviolet and visible light irradiation have also been investigated [38]. The effect of different parameters on degradation efficiency such as catalyst type and loading rate (50–500 $mg.L^{-1}$), initial drug concentration (10, 40, 80 $mg.L^{-1}$), and radiation wavelength (200–600 nm) were investigated. The results showed that exposure to visible light ($\lambda_{exc} > 390$ nm) caused a complete photocatalytic degradation reaction of both compounds. Regardless of other parameters such as the type of photocatalyst, the initial drug concentration, and the wavelength of the visible light emitted, the IBP conversion reaction rate is higher than that of CBZ. The presence of isopropanol also showed a significant inhibitory effect on CBZ degradation, which is considered as evidence of the effect of the solution phase composition [76].

An introduction to photocatalysis 17

1.4.4.6 Pesticides

Pesticides and toxicants are major pollutants in the agro-chemical industry. In one study by Alalam *et al.* (2015), the removal of pesticides from industrial wastewater was targeted. For this, nano TiO_2 was employed, and the monitored pollutants in the samples were chlorpyrifos, lambda-cyhalothrin, and diazinon. COD was also measured as a parameter regarding the level of pollution in the wastewater. The independent variables of initial pH, irradiation time, and chemical dose were altered to observe their effects on the dependent variable. The maximum removal of COD with the photo-Fenton process was 90.7%, while with photocatalytic treatment it was 79.6% [77].

Elsewhere, Kushniarou *et al.* (2019) conducted a study to photocatalytically degrade 12 conventional pesticides for vegetables, grapes, citrus fruits, and fruit crops in an aqueous suspension of TiO_2 and $Na_2S_2O_8$ in a semi-industrial unit under natural sunlight in Murcia, Spain. The optical decomposition of all pesticides can be modeled by assuming quasi-first-order kinetics. The time required to remove 90% of the contaminants in the summer was reported to be less than 4 hours, except for cyproconazole which was reported to be 4.9 hours [78].

1.5 PHOTOCATALYTIC MATERIALS

1.5.1 Semiconductors

Semiconductors are materials of choice for photocatalysts, particularly metal oxides and metal sulfides. As stated earlier, the performances of photocatalysts are directly related to their electronic band structures. To be efficient in the visible fraction of the solar spectrum, a photocatalyst must have a band gap between 3 and 1.5 eV. Photocatalysts having band gaps between 3 and 4 eV are active in the UV range. The energy level of the conduction and valence bands also plays a crucial role. In order to generate OH· radicals the valence band should have an oxidation potential above 2.31 V/ENH (potential of the couple $OH·/H^+$, H_2O), and/or the conduction band below 0.92 V/ENH (potential of the oxygen reduction in $O_2·^-$) [6]. As presented in Figure 1.6, many semiconductors have the required properties, like ZnO [79], WO_3 [80], Fe_2O_3 [81], TiO_2 [82], SnO_2 [83], CdS [84] and CdSe [85]. With a band gap of 3.8 eV, SnO_2 is active too far in the UV range, below 325 nm. WO_3, α-Fe_2O_3, CdS and CdSe are materials active in the visible range, with band gaps between 1.7 and 2.7 eV. However, the valence band energy of CdS and CdSe does not allow the oxidation of water into OH·. Concerning α-Fe_2O_3, its small band gap and the favorable edge position of its electronic bands make it a promising material for photocatalytic degradations. However, the photocatalytic performances of a-Fe_2O_3 are limited due to the quick recombination rate of e^-/h^+ carriers and the low diffusion length of h^+ (2–4 nm) [86]. The same limitations are observed for WO_3 [87].

ZnO and TiO_2 have an identical band gap around 3.3 eV. They are the two most used and studied oxides in the literature concerning the photocatalytic degradation of pollutants in water. Indeed, they are the two most performing and reliable materials in photocatalytic water treatment. They are chemically and thermally stable, not expensive and can be synthesized easily under different forms such as nanoparticles, nanowires, thin films and so on. The position of

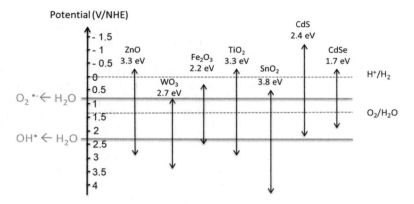

Figure 1.6 Band gaps and band positions of different semiconductors. Reprinted with permission from [4].

their bands as well as their long carrier lifetime and diffusion lengths allow them to efficiently oxidize water into OH·. Yet, these materials show photocatalytic activity in the UV range only, below 375 nm. Therefore, much effort has been made in preparing visible light active photocatalysts, by tuning ZnO and TiO_2 band structures. Reported strategies include doping with metallic (like Fe^{2+}, Co^{2+}, Ag^+, Cu^{2+}, and Mn^{2+}) [88] or nonmetallic elements (like N, C, S or P) [89], surface plasmon resonance [90], sensitization with dyes or association with other semiconductors like heterostructures or Z-schemes [91, 92].

1.5.2 Heterostructures

Heterostructures or heterojunctions created between different semiconductors have been studied for their ability to enhance light absorption in the visible range. Additionally, heterostructures have increased charge carrier lifetime and mobility with respect to the original semiconductors constituting the heterostructure [93]. A heterostructure is an interfacial association of two or more components. Depending on the band alignment of the semiconductors, three types of heterostructures can be created, as presented in Figure 1.7: type I (symmetric), type II (staggered) and type III (broken). Type I heterostructures promote the recombination of photogenerated electrons/holes. Consequently, they are often found in LED systems, where the radiative recombination of charge carriers must be maximized [94]. In the photocatalytic processes, type II heterostructures are particularly attractive, since they promote the separation and delocalization of photogenerated charge carriers. Thanks to a built-in electric potential created at the interface of the two semiconductors, holes are driven in the valence band (VB) of one semiconductor, while electrons are driven in the conduction band (CB) of the second one [95]. Subsequently, photogenerated charge carrier lifetimes are increased. Type III or broken heterostructures find applications in tunneling field effect transistors [96].

An introduction to photocatalysis 19

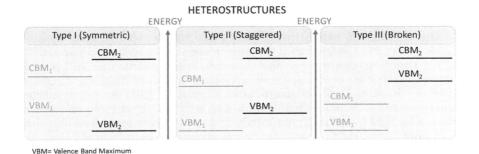

Figure 1.7 Schematic representation of the three possible types of heterostructures. Reproduced from [97].

In water treatment or hydrogen production applications, type II heterostructures between two semiconductors have been particularly investigated in the literature. This kind of junction can be created between different metal oxides, metal sulfides or carbon-based structures such as, for example, ZnO [98], TiO$_2$ [99], SnO$_2$ [100], CdS [101], WO$_3$ [102], and g-C$_3$N$_4$ [103]. In the case of a heterostructure like ZnO/SnO$_2$ (Figure 1.8), the improvement of photocatalytic properties is due to longer lifetime of charge carriers, induced by the electron/hole separation towards the opposite sides of the junction [100]: electrons are driven from the valence band of ZnO to the valence band of SnO$_2$, and the holes are driven from the conduction band of SnO$_2$ to the conduction band of ZnO. Thus, the photocatalytic degradation kinetic of methylene blue is quicker in the presence of the heterostructure. In

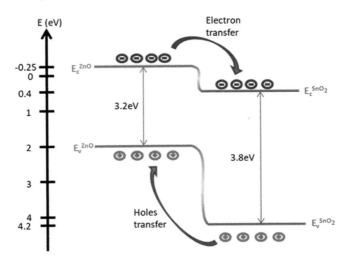

Figure 1.8 Schematic representation of charge carrier behavior in type II heterostructure ZnO/SnO$_2$. Reproduced from [20].

addition, thanks to its high chemical and thermal stability, SnO_2 acts also as a protective material against the degradation/dissolution of ZnO in water [97].

Other kinds of heterostructures attempt to use visible light sensitive semiconductors in order to shift the absorbance and the photocatalytic activity of the structure toward the visible range. Such heterostructures still provide efficient separation of charge carriers. Good examples of these are ZnO/CuS [104] or TiO_2/CdS [105] heterostructures. Basu *et al.* [104] demonstrated that ZnO nanotubes decorated with a CuS shell were 2.5 times more efficient for the visible light (>400 nm) photocatalytic degradation of methylene blue than CuS only. In the visible range, ZnO alone showed no activity at all. Here, the double effect of the heterostructure is highlighted: photocatalysis efficiency improvement and visible light activity.

1.5.3 Z-Scheme heterostructures

A fourth type of heterostructure, possessing similar band alignments to the type II heterostructure, is the Z-scheme heterostructure. In the case of a Z-scheme junction, the charge carriers' migration mechanism is different from that of the type II heterostructure. Holes generated in the valence band of semiconductor 'A' recombine with electrons generated in the conduction band of semiconductor 'B' as shown in Figure 1.9. Consequently, holes with the highest oxidative potential (lowest valence band) and electrons with the highest reductive potential (highest conduction band) remain available for the photocatalytic action, unlike in the type II heterostructure, where electrons and holes are driven to lower energy states.

Three different kinds of Z-schemes can be developed: the liquid phase Z-scheme, the all-solid-state Z-scheme and the direct Z-scheme (Figure 1.10). The liquid phase Z-scheme was first discovered in 1979 [107]. It is prepared by associating two different semiconductors with a shuttle redox mediator, via an electron acceptor/donor (A/D) pair like Fe^{2+}/Fe^{3+} or I^-/IO^{3-} [108]. The all-solid-state Z-scheme, the second technology developed, relies on the use of a solid electron mediator between the two semiconductors. The electron

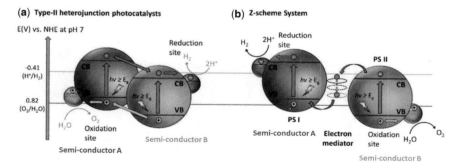

Figure 1.9 Schematic representation of carrier recombination in (a) type II heterojunction, (b) Z-scheme heterojunction. Reproduced from [106].

An introduction to photocatalysis

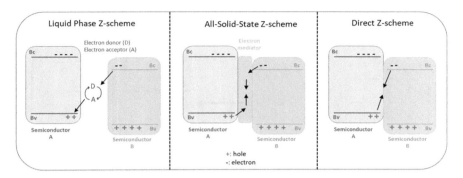

Figure 1.10 Schematic representation of the three different types of Z-scheme: the liquid phase Z-scheme, the all-solid-state Z-scheme and the direct Z-scheme.

mediator is typically a noble metal (Au, Pt, Ag) in the form of nanoparticles or a thin film, or a conductive carbon structure like graphene sheets, nanotubes or quantum dots [109–111]. More recently, the third generation of Z-scheme, the direct Z-scheme, consists of semiconductors that are both directly in contact, like in the type II heterostructure.

The difference between a type II heterostructure and a direct Z-scheme results from the electronic configuration determined by the Fermi level of the two semiconductors to couple (Figure 1.11). Graphitic carbon nitride g-C$_3$N$_4$ has a higher Fermi level than ZnO [108]. Thus, when the two materials contact, electrons from the g-C$_3$N$_4$ migrate toward the ZnO during the Fermi level alignment at the interface. The local electron depletion induces an upward band bending in the g-C$_3$N$_4$ and a downward band bending in the ZnO. As the conduction band minimum and valence band maximum of g-C$_3$N$_4$ are higher than those of ZnO, it results in the formation of a direct Z-scheme between g-C$_3$N$_4$ and ZnO. In the case of a heterojunction based on ZnO and SnO$_2$, the Fermi level of SnO$_2$ being higher than that of ZnO [112] leads to an upward band bending in the SnO$_2$ and a downward band bending in the ZnO. As the conduction band minimum and valence band maximum of ZnO are higher than those of SnO$_2$, the band bending leads to the formation of a type II heterostructure.

Huang et al. [113] showed that between the two semiconductors g-C$_3$N$_4$ and W$_{18}$O$_{49}$, a type II heterostructure or a direct Z-scheme could be obtained depending on their band bending at the interface. When the two bare materials form a junction, the Fermi level of W$_{18}$O$_{49}$ is higher than that of g-C$_3$N$_4$ leading to a type II heterostructure. However, if triethanolamine is adsorbed at the surface of those two materials, the Fermi level of g-C$_3$N$_4$ is upshifted above the Fermi level of W$_{18}$O$_{49}$, forming a direct Z-scheme junction.

Z-scheme photocatalysts are considered as a promising technological solution for visible light driven treatment of water. Indeed, just like for type II heterostructures, the association of a visible light sensitive semiconductor with an efficient photocatalyst like ZnO or TiO$_2$ provides photocatalysts

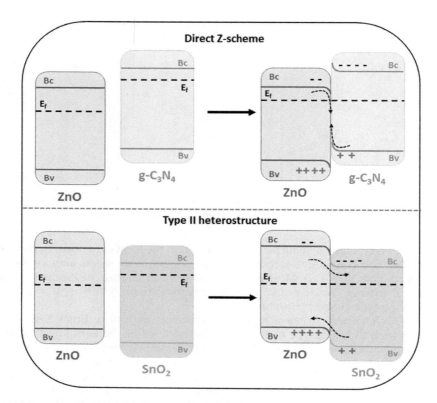

Figure 1.11 Schematic representation of the formation of type II heterostructure and direct Z-scheme between ZnO/SnO₂ and ZnO/g-C₃N₄ depending on their Fermi level energy difference.

with the possibility of working under solar light. On top of that, Z-scheme photocatalysts have an advantage over type II heterostructures, as they promote the preservation of high energy carriers in the lowest valence band and the highest conduction band. Therefore, Z-scheme photocatalysts like $TiO_2/g-C_3N_4$ [114], $ZnO/g-C_3N_4$ [115], $ZnIn_2S_4/Bi_2WO_6$ [116] or ZnO/CdS [117] have been thoroughly studied in order to prove their efficiency for photocatalytic water purification as well as for hydrogen production by water splitting.

1.6 OPERATIONAL PARAMETERS AFFECTING THE PHOTOCATALYTIC PROCESS

There are various parameters that can change the oxidation reaction rate and the efficiency of photocatalytic processes. In the following, the effect of several operating parameters are reviewed.

1.6.1 Effect of pH

One of the important parameters affecting photocatalytic processes is the effect of pH. The pH of the solution affects the degree of ionization, agglomeration, oxidation potential of the photocatalytic capacity band, and the absorption of pollutants [118]. Surface charge is one of the important parameters in the photocatalytic process that is affected by pH. The pH value at which the surface charge is completely neutralized is known as the point of zero charge (PZC). In many studies, the PZC of titanium dioxide has been used to investigate the effect of pH on its photocatalytic oxidation performance. Titanium dioxide photocatalysts have shown a PZC of 4.5 to 7, depending on the type and composition. At the PZC point, due to the lack of electrostatic force, the attraction between contaminants in water and photocatalytic particles is minimal. At pH less than PZC the surface of the photocatalyst is charged with positive charges and generates electrostatic attraction for negative charges and vice versa. The following shows the reactions of titanium dioxide with respect to the PZC [119]:

$$pH < pH_{PZC} \quad TiOH + H^+ \rightarrow TiOH_2^+ \tag{1.18}$$

$$pH > pH_{PZC} \quad TiOH + OH^- \rightarrow TiO^- + H_2O \tag{1.19}$$

Sun $et\ al.$ (2019) controlled the morphology of a $BiVO_4$ catalyst by adjusting the pH of the solution to degrade phenol-containing effluent. The results showed that the photocatalytic activity of $BiVO_4$ for phenol degradation was improved by increasing the pH value. Reasons for higher photocatalytic activity include altered morphology, higher ability to absorb sunlight, smaller band gaps, and less recombination of electron pairs [120].

In another study, Guo $et\ al.$ (2020) investigated the photocatalytic activity of $AgBr/Ag_2CO_3$ synthesized by the in situ growth method at different pHs. The results showed that the $AgBr/Ag_2CO_3$ heterostructure had a higher photocatalytic activity than pure AgBr and Ag_2CO_3. In addition, different pH conditions (between pH 7 to 10) caused changes in photocatalytic activity due to their effect on crystallization. The $AgBr/Ag_2CO_3$ photocatalyst produced at $pH=9$ had the highest photocatalytic activity and the degradation rate of rhodamine B with an initial concentration of $5\ mg.L^{-1}$ in 20 minutes was 98.93% [121].

Intaphong $et\ al.$ (2020) investigated the effect of pH on the crystal structure, morphology and photocatalytic behavior of BiOBr under visible light. The photocatalytic properties of BiOBr with different morphologies made using the hydrothermal method were investigated for optical decomposition of rhodamine B. Hierarchical micro-flowers at $pH=8$ showed the best photocatalytic activity with a decolorization efficiency of nearly 98% [122].

1.6.2 Temperature

Another important parameter that has a great impact on photocatalytic activity and degradation of pollutants is temperature. Most photocatalytic reactions take place at room temperature, however, at temperatures below

0°C, the desorption rate of the final product decreases and thus increases the activation energy. Also, by increasing the reaction temperature above 80°C, the adsorption of the reactant becomes a limiting factor [123].

Ariza-Tarazona *et al.* (2020) investigated the effect of temperature and pH on the photocatalytic degradation of microplastic contaminants with the help of carbon and nitrogen-doped titanium dioxide photocatalyst. The concentration of microplastic contaminants in the experiments was $4000 \ mg.L^{-1}$. Here, low temperature (0°C) increased the surface of microplastic by means fragmentation and low pH (pH $= 3$) caused the formation of hydroperoxide during photooxidation [124].

Neto *et al.* (2019) investigated the photocatalytic effect of Ag_2WO_4 nanorods at different temperatures (10, 30, 50, 70, and 90°C) on the degradation of methylene blue (MB) and methyl orange (MO). The increase in temperature lengthened and reduced the thickness of the nanorods, thereby increasing their surface area, and their rate of absorption. Each photocatalyst was used in four photocatalytic cycles to accurately evaluate performance and stability [125].

1.6.3 Presence of oxidants

Oxidants (e.g. H_2O_2, $KBrO_3$, and HNO_3) are external agents that are added to the reaction as irreversible electron acceptors to help produce intermediate radicals to remove contaminants. Therefore, oxidants are electron scavengers from the valance band and increase the efficiency of the photocatalytic process by: (1) reduction of electron/hole pair recombination time, (2) increased production of OH· to destroy pollutants, and (3) production of oxidant species for increasing oxidation rates [126].

1.6.4 Pollutant concentration

Studying the relationship between pollutant concentration and the photocatalytic degradation rate is of great importance for the practical design of photocatalytic treatment units. The degradation reaction of many contaminants follows pseudo-first-order kinetics that can be corrected in the form of the Langmuir-Hinshelwood equation for solid-liquid reactions.

$$\ln\left(C_0/C\right) = k_r Kt = k_1 t \tag{1.20}$$

In this equation, k_1 is the first-order reaction constant, t is the time required to reduce the concentration from the initial concentration (C_0) to the final concentration (C), K is the equilibrium constant for adsorption of the contaminant to the catalyst surface, and k_r is the reaction limiting rate [127].

In the case of color compounds, the amount of degradation may increase with increasing color concentration (because more molecules will be available for degradation), but after reaching a certain critical concentration, this amount begins to decrease. This reduction can be attributed to the reduction in the amount of ultraviolet radiation reaching the photocatalyst surface. In most studies, a concentration of pollutants in the range of $10–200 \ mg.L^{-1}$ has been used, which corresponds to the amount in most real wastewater samples [128].

An introduction to photocatalysis

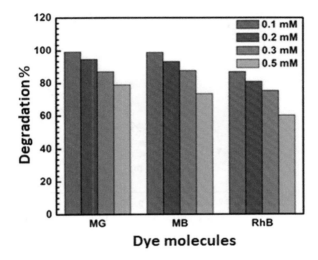

Figure 1.12 The relationship between initial pollutant concentration and photocatalytic degradation of dyes. Reprinted from [129].

Boruah *et al.* (2016) investigated the photocatalytic degradation of methyl green (MG), methyl blue (MB), and rhodamine B (RhB), and the reduction of Cr (VI) in an aqueous medium using a recyclable magnetic catalyst of Fe_3O_4/reduced-graphene-oxide (rGO) under visible light. The effect of initial pigment concentration on the rate of photocatalytic degradation was one of the important parameters investigated in this study. At a constant concentration of 0.5 g.L^{-1} of photocatalyst and pH = 5, the concentration of pigment molecules was altered between 0.08 and 0.5 mM. Figure 1.12 shows that at an initial concentration of 0.1 mM, the rate of photocatalytic degradation of all three pigments is over 98%. However, the efficiency of photocatalytic degradation decreased with increasing initial photocatalyst concentration [129]. This implies that even the lowest concentration of dye might have hindered the penetration of adequate light into the solution.

Elsewhere, Chanu *et al.* (2019) investigated the effect of operating parameters on the photocatalytic degradation of methylene blue pigment by manganese doped zinc oxide (ZnO) nanoparticles. To investigate the effect of the initial concentration of the pigment, a constant concentration of catalyst (0.30 g.L^{-1}) and pH = 12 were used with varying pigment concentration. The graph of $\ln(C_0/C)$ versus radiation (Figure 1.13) shows the maximum amount of photocatalytic degradation at a concentration of 10 ppm. Increased degradation with increasing concentration has been attributed to increasing the probability of the dye molecules colliding with the OH· radical. Also, the reduction of the reaction rate after the optimal concentration of contaminants has been attributed to the coverage of the catalyst surface by contaminants and the reduction of the produced OH· radical [130].

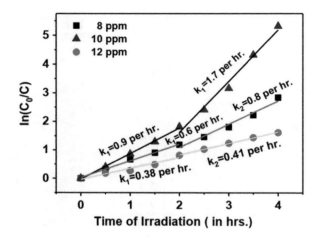

Figure 1.13 The graph of ln(C_0/C) versus time at different concentrations of dye. Reprinted from [130].

1.6.5 Catalyst loading

Increasing the amount of catalyst will logically increase active sites in the solution, resulting in more photon adsorption and greater production of OH· hydroxyl radicals and irradiated positive holes. However, an excessive increase in the amount of catalyst can reduce the amount of photocatalytic activity. One of the important reasons for this decrease in activity is the increase in solution turbidity and light scattering, which reduces the number of photons absorbed by the photocatalyst [131, 132].

Paul *et al.* (2019) investigated the effect of operational parameters on photocatalytic degradation of methylene blue (MB) by urea-based graphite nitride carbonate (g-C_3N_4). The results showed that increasing the amount of photocatalyst from 0.01 g to 0.05 g caused an increase in photocatalyst activity, which was attributed to the increase in active sites. Increasing the amount of photocatalyst beyond 0.05 g reduced the activity of the photocatalyst, and this has been attributed to the reduction of light received by the photocatalyst. Figure 1.14 shows the concentration of methylene blue relative to its initial concentration with respect to time [133].

1.6.6 Light intensity and wavelength

As stated earlier, in order for light to activate a photocatalytic substance, its energy must be at least equal to the energy of the photocatalyst band gap [132]. Creating charge carriers helps produce free radicals to destroy pollutants. Therefore, the amount of degradation is affected by the intensity of light, and the distribution of light in a photocatalytic reactor determines the efficiency of the pollutant conversion and the amount of degradation. In many studies, the amount of pollutant degradation has been shown to depend

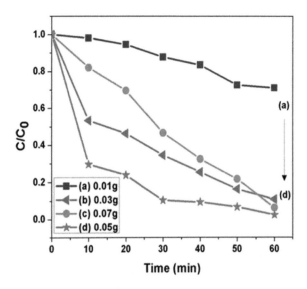

Figure 1.14 Comparison of photocatalytic activity of g-C_3N_4 with different amounts of photocatalyst loading. Initial MB concentration is 10 ppm and pH = 11. Reprinted from [133].

linearly on the intensity of the light, while in others, the relationship between the amount of degradation and the square of the light intensity is linear. It should be noted that at high intensities the reaction rate is independent of light intensity [134].

For example, in one study by Chen *et al.* (2019), carbon quantum dots were modified by $K_2Ti_6O_{13}$ nanotubes to produce a composite. It was hypothesized that this modification would help the degradation of amoxicillin under visible light. For this irradiation under wavelengths ranging from 420 to 630 nm were investigated. With increasing wavelength, the photocatalytic activity decreased due to changes in photon energy. Furthermore, the composite performed significantly more effectively compared with when $K_2Ti_6O_{13}$ was used alone[135].

Elsewhere, Rahimi Aghdam *et al.* (2018) investigated the removal of NOX by stabilized $BiFeO_3$ photocatalytic nanoparticles and the operating parameters affecting this process. In this study, $BiFeO_3$ perovskite (BFO) was synthesized by the sol-gel auto-combustion method. The gap band energy was calculated to be about 2.13 eV and the specific surface area of the prepared BFO nanostructure was 55.1 $m^2.g^{-1}$. The effect of UV radiation power from two UV lamps, 8 and 15 W, was investigated. As can be seen in Figure 1.15, the percentage of NO conversion was higher in the presence of the 15 W lamp. Logically, the higher ultraviolet light intensity can produce more electron/hole pairs, which accelerates the optical decomposition [136].

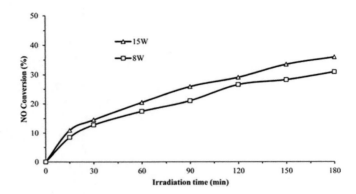

Figure 1.15 The effect of light power on photocatalytic removal performance with initial NO concentration of 5 ppm. Reprinted from [136].

1.7 CONCLUSION

This chapter has been dedicated to the introduction of photocatalytic principles and processes involved in water treatment and energy production (hydrogen/oxygen through water splitting). Mechanisms implicated in the photogeneration of charge carriers in semiconductors and charge carrier activities toward organic pollutant degradation or water splitting have been detailed. The application of photocatalytic methods to degrade multiple molecules such as dyes, phenols, nitrogen-containing molecules, pesticides and pharmaceuticals has been studied, along with characterization techniques that are able to display charge carrier lifetimes and spatial distribution in photocatalytic materials. Also, the main materials showing photocatalytic properties have been highlighted, as well as the different strategies considered to improve their photocatalytic activity. Particularly, the formation of the different heterostructures formed between semiconductors, or experimental parameters affecting the kinetic of reactions have been described.

REFERENCES

1. Ameta SC. Chapter 1 - Introduction. In: Ameta SC, Ameta R, eds. *Advanced Oxidation Processes for Waste Water Treatment*: Academic Press, Cambridge, Massachusetts, United States, 2018: 1–12.
2. Ebrahiem EE, Al-Maghrabi MN, Mobarki AR. Removal of organic pollutants from industrial wastewater by applying photo-Fenton oxidation technology. *Arabian Journal of Chemistry* 2017; **10**: S1674–S9.
3. Clarizia L, Russo D, Di Somma I, Marotta R, Andreozzi R. Homogeneous photo-Fenton processes at near neutral pH: A review. *Applied Catalysis B: Environmental* 2017; **209**: 358–71.
4. Rogé V. Etude, fabrication et caractérisation de nanostructures catalytiques de type ZnO/SnO$_2$ intégrées à des membranes modèles pour la dépollution de l'eau: Université de Strasbourg, 2015.
5. Rogé V, Bahlawane N, Lamblin G, *et al.* Improvement of the photocatalytic degradation property of atomic layer deposited ZnO thin films: the interplay

between film properties and functional performances. *Journal of Materials Chemistry A* 2015; **3**: 11453–61.

6. Buettner GR. The Pecking Order of Free Radicals and Antioxidants Lipid Peroxidation, α-Tocopherol, and Ascorbate. *Archives of Biochemistry and Biophysics* 1993; **300**: 535–43.
7. Maeda K. Photocatalytic water splitting using semiconductor particles: History and recent developments. *Journal of Photochemistry and Photobiology C: Photochemistry Reviews* 2011; **12**: 237–68.
8. Maeda K, Domen K. Photocatalytic Water Splitting: Recent Progress and Future Challenges. *The Journal of Physical Chemistry Letters* 2010; **1**: 2655–61.
9. Takanabe K. Photocatalytic Water Splitting: Quantitative Approaches toward Photocatalyst by Design. *ACS Catalysis* 2017; **7**: 8006–22.
10. Foglia L, Vempati S, Tanda Bonkano B, *et al.* Revealing the competing contributions of charge carriers, excitons, and defects to the non-equilibrium optical properties of ZnO. *Struct Dyn* 2019; **6**: 034501.
11. Gao Y, Nie W, Wang X, Fan F, Li C. Advanced space- and time-resolved techniques for photocatalyst studies. *Chem Commun (Camb)* 2020; **56**: 1007–21.
12. Berera R, van Grondelle R, Kennis JT. Ultrafast transient absorption spectroscopy: principles and application to photosynthetic systems. *Photosynth Res* 2009; **101**: 105–18.
13. Nakajima S, Katoh R. Time-resolved microwave conductivity study of charge carrier dynamics in commercially available TiO_2 photocatalysts. *Journal of Materials Chemistry A* 2015; **3**: 15466–72.
14. C. Colbeau-Justin MAV. Time-resolved microwave conductivity (TRMC) a useful characterization tool for charge carriers transfert in photocatalysis-a short review. *Revista Mexicana de Fisica* 2013; **59**: 191–200.
15. Tahiri Alaoui O, Herissan A, Le Quoc C, *et al.* Elaboration, charge-carrier lifetimes and activity of Pd-TiO_2 photocatalysts obtained by gamma radiolysis. *Journal of Photochemistry and Photobiology A: Chemistry* 2012; **242**: 34–43.
16. Brüninghoff R, Wenderich K, Korterik JP, Mei BT, Mul G, Huijser A. Time-Dependent Photoluminescence of Nanostructured Anatase TiO_2 and the Role of Bulk and Surface Processes. *The Journal of Physical Chemistry C* 2019; **123**: 26653–61.
17. Xing J, Chen ZP, Xiao FY, *et al.* Cu-Cu_2O-TiO_2 nanojunction systems with an unusual electron-hole transportation pathway and enhanced photocatalytic properties. *Chem Asian J* 2013; **8**: 1265–70.
18. Dillon RJ, Joo JB, Zaera F, Yin Y, Bardeen CJ. Correlating the excited state relaxation dynamics as measured by photoluminescence and transient absorption with the photocatalytic activity of Au@TiO_2 core-shell nanostructures. *Phys Chem Chem Phys* 2013; **15**: 1488–96.
19. Godin R, Wang Y, Zwijnenburg MA, Tang J, Durrant JR. Time-Resolved Spectroscopic Investigation of Charge Trapping in Carbon Nitrides Photocatalysts for Hydrogen Generation. *J Am Chem Soc* 2017; **139**: 5216–24.
20. Rogé V, Georgantzopoulou A, Mehennaoui K, *et al.* Tailoring the optical properties of ZnO nano-layers and their effect on in vitro biocompatibility. *RSC Advances* 2015; **5**: 97635–47.
21. Zhang Y, Zhang Y, Song L, *et al.* Illustration of charge transfer in graphene-coated hexagonal ZnO photocatalysts using Kelvin probe force microscopy. *RSC Advances* 2018; **8**: 885–94.
22. Zhu J, Fan F, Chen R, An H, Feng Z, Li C. Direct Imaging of Highly Anisotropic Photogenerated Charge Separations on Different Facets of a Single $BiVO_4$ Photocatalyst. *Angew Chem Int Ed Engl* 2015; **54**: 9111–4.

23. Jian A, Feng K, Jia H, Zhang Q, Sang S, Zhang X. Quantitative investigation of plasmonic hot-electron injection by KPFM. *Applied Surface Science* 2019; **492**: 644–50.
24. Marrese M, Guarino V, Ambrosio L. Atomic Force Microscopy: A Powerful Tool to Address Scaffold Design in Tissue Engineering. *J Funct Biomater* 2017; **8**(1): 7.
25. Yu W, Fu HJ, Mueller T, Brunschwig BS, Lewis NS. Atomic force microscopy: Emerging illuminated and operando techniques for solar fuel research. *J Chem Phys* 2020; **153**: 020902.
26. Eichhorn J, Kastl C, Cooper JK, *et al.* Nanoscale imaging of charge carrier transport in water splitting photoanodes. *Nat Commun* 2018; **9**: 2597.
27. Viswanathan B. Photocatalytic Degradation of Dyes: An Overview. *Current Catalysis* 2018; **7**: 99–121.
28. Rogé V, Guignard C, Lamblin G, *et al.* Photocatalytic degradation behavior of multiple xenobiotics using MOCVD synthesized ZnO nanowires. *Catalysis Today* 2018; **306**: 215–22.
29. Dashairya L, Mehta A, Saha P, Basu S. Visible-light-induced enhanced photocatalytic degradation of Rhodamine-B dye using BixSb2-xS3 solid-solution photocatalysts. *J Colloid Interface Sci* 2020; **561**: 71–82.
30. Lee SY, Kang D, Jeong S, Do HT, Kim JH. Photocatalytic Degradation of Rhodamine B Dye by TiO_2 and Gold Nanoparticles Supported on a Floating Porous Polydimethylsiloxane Sponge under Ultraviolet and Visible Light Irradiation. *ACS Omega* 2020; **5**: 4233–41.
31. Bayomie OS, Kandeel H, Shoeib T, Yang H, Youssef N, El-Sayed MMH. Novel approach for effective removal of methylene blue dye from water using fava bean peel waste. *Sci Rep* 2020; **10**: 7824.
32. Li M, Guan R, Li J, *et al.* Photocatalytic Performance and Mechanism Research of Ag/HSTiO2 on Degradation of Methyl Orange. *ACS Omega* 2020; **5**: 21451–7.
33. Bahrudin NN, Nawi MA, Nawawi WI. Enhanced photocatalytic decolorization of methyl orange dye and its mineralization pathway by immobilized TiO2/polyaniline. *Research on Chemical Intermediates* 2019; **45**: 2771–95.
34. Pino E, Calderon C, Herrera F, Cifuentes G, Arteaga G. Photocatalytic Degradation of Aqueous Rhodamine 6G Using Supported TiO_2 Catalysts. A Model for the Removal of Organic Contaminants From Aqueous Samples. *Front Chem* 2020; **8**: 365.
35. Hu C, Yu JC, Hao Z, Wong PK. Photocatalytic degradation of triazine-containing azo dyes in aqueous TiO_2 suspensions. *Applied Catalysis B: Environmental* 2003; **42**: 47–55.
36. Konstantinou IK, Albanis TA. TiO_2-assisted photocatalytic degradation of azo dyes in aqueous solution: kinetic and mechanistic investigations. *Applied Catalysis B: Environmental* 2004; **49**: 1–14.
37. Divya N, Bansal A, Jana AK. Photocatalytic degradation of azo dye Orange II in aqueous solutions using copper-impregnated titania. *International Journal of Environmental Science and Technology* 2013; **10**: 1265–74.
38. Islam MR, Islam JB, Furukawa M, Tateishi I, Katsumata H, Kaneco S. Photocatalytic Degradation of a Systemic Herbicide: Picloram from Aqueous Solution Using Titanium Oxide (TiO_2) under Sunlight. *ChemEngineering* 2020; **4**(4): 58.
39. Nguyen TT, Nam SN, Kim J, Oh J. Photocatalytic degradation of dissolved organic matter under ZnO-catalyzed artificial sunlight irradiation system. *Sci Rep* 2020; **10**: 13090.
40. Balcioglu IA, Arslan I, Sacan MT. Homogenous and Heterogenous Advanced Oxidation of Two Commercial Reactive Dyes. *Environmental Technology* 2001; **22**: 813–22.

41. Fenton HJH. LXXIII.–Oxidation of tartaric acid in presence of iron. *Journal of the Chemical Society, Transactions* 1894; **65**: 899–910.
42. Barbot E, Vidic NS, Gregory KB, Vidic RD. Spatial and Temporal Correlation of Water Quality Parameters of Produced Waters from Devonian-Age Shale following Hydraulic Fracturing. *Environmental Science & Technology* 2013; **47**: 2562–9.
43. Walling C, Goosen A. Mechanism of the ferric ion catalyzed decomposition of hydrogen peroxide. Effect of organic substrates. *Journal of the American Chemical Society* 1973; **95**: 2987–91.
44. Ahile UJ, Wuana RA, Itodo AU, Sha'Ato R, Malvestiti JA, Dantas RF. Are iron chelates suitable to perform photo-Fenton at neutral pH for secondary effluent treatment? *Journal of Environmental Management* 2021; **278**: 111566.
45. Ahmed SN, Haider W. Heterogeneous photocatalysis and its potential applications in water and wastewater treatment: a review. *Nanotechnology* 2018; **29**: 342001.
46. Malato S, Fernández-Ibáñez P, Maldonado MI, Blanco J, Gernjak W. Decontamination and disinfection of water by solar photocatalysis: Recent overview and trends. *Catalysis Today* 2009; **147**: 1–59.
47. Biancullo F, Moreira NFF, Ribeiro AR, *et al.* Heterogeneous photocatalysis using UVA-LEDs for the removal of antibiotics and antibiotic resistant bacteria from urban wastewater treatment plant effluents. *Chemical Engineering Journal* 2019; **367**: 304–13.
48. Ayekoe CYP, Robert D, Lanciné DG. Combination of coagulation-flocculation and heterogeneous photocatalysis for improving the removal of humic substances in real treated water from Agbô River (Ivory-Coast). *Catalysis Today* 2017; **281**: 2–13.
49. Kawawaki T, Mori Y, Wakamatsu K, *et al.* Controlled colloidal metal nanoparticles and nanoclusters: recent applications as cocatalysts for improving photocatalytic water-splitting activity. *Journal of Materials Chemistry A* 2020; **8**: 16081–113.
50. Luo H, Zeng Z, Zeng G, *et al.* Recent progress on metal-organic frameworks based-and derived-photocatalysts for water splitting. *Chemical Engineering Journal* 2020; **383**: 123196.
51. Shi W, Li M, Huang X, Ren H, Yan C, Guo F. Facile synthesis of 2D/2D $Co_3(PO_4)_2$/g-C_3N_4 heterojunction for highly photocatalytic overall water splitting under visible light. *Chemical Engineering Journal* 2020; **382**: 122960.
52. Li D, Kassymova M, Cai X, Zang S-Q, Jiang H-L. Photocatalytic CO_2 reduction over metal-organic framework-based materials. *Coordination Chemistry Reviews* 2020; **412**: 213262.
53. White JL, Baruch MF, Pander JE, *et al.* Light-Driven Heterogeneous Reduction of Carbon Dioxide: Photocatalysts and Photoelectrodes. *Chemical Reviews* 2015; **115**: 12888–935.
54. Wang Y, Wang S, Zhang SL, Lou XW. Formation of Hierarchical $FeCoS_2$–CoS_2 Double-Shelled Nanotubes with Enhanced Performance for Photocatalytic Reduction of CO_2. *Angewandte Chemie International Edition* 2020; **59**: 11918–22.
55. Vasilachi IC, Asiminicesei DM, Fertu DI, Gavrilescu M. Occurrence and Fate of Emerging Pollutants in Water Environment and Options for Their Removal. *Water* 2021; **13**(2): 181.
56. Ameta R, Ameta SC. *Photocatalysis: Principles and Applications*: CRC Press LLC, Boca Raton, Florida, United States, 2019.
57. Han F, Kambala VSR, Srinivasan M, Rajarathnam D, Naidu R. Tailored titanium dioxide photocatalysts for the degradation of organic dyes in wastewater treatment: A review. *Applied Catalysis A: General* 2009; **359**: 25–40.

58. Houas A, Lachheb H, Ksibi M, Elaloui E, Guillard C, Herrmann J-M. Photocatalytic degradation pathway of methylene blue in water. *Applied Catalysis B: Environmental* 2001; **31**: 145–57.
59. Arslan I, Balcioğlu IA. Degradation of commercial reactive dyestuffs by heterogenous and homogenous advanced oxidation processes: a comparative study. *Dyes and Pigments* 1999; **43**: 95–108.
60. Weber EJ, Adams RL. Chemical- and Sediment-Mediated Reduction of the Azo Dye Disperse Blue 79. *Environmental Science & Technology* 1995; **29**: 1163–70.
61. Qi X-H, Zhuang Y-Y, Yuan Y-C, Gu W-X. Decomposition of aniline in supercritical water. *Journal of Hazardous Materials* 2002; **90**: 51–62.
62. Kansal SK, Kaur N, Singh S. Photocatalytic Degradation of Two Commercial Reactive Dyes in Aqueous Phase Using Nanophotocatalysts. *Nanoscale Research Letters* 2009; **4**: 709.
63. Eriksson E, Baun A, Mikkelsen PS, Ledin A. Risk assessment of xenobiotics in stormwater discharged to Harrestrup Å, Denmark. *Desalination* 2007; **215**: 187–97.
64. Ahmed S, Rasul MG, Martens WN, Brown R, Hashib MA. Heterogeneous photocatalytic degradation of phenols in wastewater: A review on current status and developments. *Desalination* 2010; **261**: 3–18.
65. Pardeshi SK, Patil AB. Solar photocatalytic degradation of resorcinol a model endocrine disrupter in water using zinc oxide. *Journal of Hazardous Materials* 2009; **163**: 403–9.
66. Lam S-M, Sin J-C, Abdullah AZ, Mohamed AR. Photocatalytic degradation of resorcinol, an endocrine disrupter, by TiO_2 and ZnO suspensions. *Environmental Technology* 2013; **34**: 1097–106.
67. Wei Z, Liang F, Liu Y, *et al.* Photoelectrocatalytic degradation of phenol-containing wastewater by TiO_2/g-C3N4 hybrid heterostructure thin film. *Applied Catalysis B: Environmental* 2017; **201**: 600–6.
68. Weiping W, Yongkui H, Shuijin Y. Photocatalytic degradation of nitrobenzene wastewater with $H_3PW_{12}O_{40}/TiO_2$ 2010 International Conference on Mechanic Automation and Control Engineering 2010: 1303–5.
69. Luo X, Chen C, Yang J, *et al.* Characterization of La/Fe/TiO_2 and Its Photocatalytic Performance in Ammonia Nitrogen Wastewater. *International Journal of Environmental Research and Public Health* 2015; **12**(11): 14626–14639.
70. Poulton SW, Krom MD, Rijn JV, Raiswell R. The use of hydrous iron (III) oxides for the removal of hydrogen sulphide in aqueous systems. *Water Research* 2002; **36**: 825–34.
71. Diya'uddeen BH, Daud WMAW, Abdul Aziz AR. Treatment technologies for petroleum refinery effluents: A review. *Process Safety and Environmental Protection* 2011; **89**: 95–105.
72. Bharatvaj J, Preethi V, Kanmani S. Hydrogen production from sulphide wastewater using Ce^{3+}–TiO_2 photocatalysis. *International Journal of Hydrogen Energy* 2018; **43**: 3935–45.
73. Rizzo L, Meric S, Guida M, Kassinos D, Belgiorno V. Heterogenous photocatalytic degradation kinetics and detoxification of an urban wastewater treatment plant effluent contaminated with pharmaceuticals. *Water Research* 2009; **43**: 4070–8.
74. Iervolino G, Zammit I, Vaiano V, Rizzo L. Limitations and Prospects for Wastewater Treatment by UV and Visible-Light-Active Heterogeneous Photocatalysis: A Critical Review. *Topics in Current Chemistry* 2019; **378**: 7.
75. Deng F, Zhao L, Luo X, Luo S, Dionysiou DD. Highly efficient visible-light photocatalytic performance of Ag/AgIn5S8 for degradation of tetracycline hydrochloride and treatment of real pharmaceutical industry wastewater. *Chemical Engineering Journal* 2018; **333**: 423–33.

An introduction to photocatalysis

76. Georgaki I, Vasilaki E, Katsarakis N. A Study on the Degradation of Carbamazepine and Ibuprofen by TiO_2 & ZnO Photocatalysis upon UV/Visible-Light Irradiation. *American Journal of Analytical Chemistry* 2014; **05**: 518–34.

77. Gar Alalm M, Tawfik A, Ookawara S. Comparison of solar TiO_2 photocatalysis and solar photo-Fenton for treatment of pesticides industry wastewater: Operational conditions, kinetics, and costs. *Journal of Water Process Engineering* 2015; **8**: 55–63.

78. Kushniarou A, Garrido I, Fenoll J, *et al.* Solar photocatalytic reclamation of agrowaste water polluted with twelve pesticides for agricultural reuse. *Chemosphere* 2019; **214**: 839–45.

79. Rodrigues J, Hatami T, Rosa JM, Tambourgi EB, Mei LHI. Photocatalytic degradation using ZnO for the treatment of RB 19 and RB 21 dyes in industrial effluents and mathematical modeling of the process. *Chemical Engineering Research and Design* 2020; **153**: 294–305.

80. Liu Y, Zeng X, Easton CD, *et al.* An in situ assembled WO_3–TiO_2 vertical heterojunction for enhanced Z-scheme photocatalytic activity. *Nanoscale* 2020; **12**: 8775–84.

81. Hitam CNC, Jalil AA. A review on exploration of Fe_2O_3 photocatalyst towards degradation of dyes and organic contaminants. *Journal of Environmental Management* 2020; **258**: 110050.

82. Ishchenko OM, Lamblin G, Guillot J, *et al.* Mesoporous TiO_2 anatase films for enhanced photocatalytic activity under UV and visible light. *RSC Advances* 2020; **10**: 38233–43.

83. Ma CM, Hong GB, Lee SC. Facile Synthesis of Tin Dioxide Nanoparticles for Photocatalytic Degradation of Congo Red Dye in Aqueous Solution. *Catalysts* 2020; **10**(7): 792.

84. Nasir JA, Rehman Zu, Shah SNA, Khan A, Butler IS, Catlow CRA. Recent developments and perspectives in CdS-based photocatalysts for water splitting. *Journal of Materials Chemistry A* 2020; **8**: 20752–80.

85. Sun C, Li T, Wen W, Luo X, Zhao L. ZnSe/CdSe core–shell nanoribbon arrays for photocatalytic applications. *CrystEngComm* 2020; **22**: 895–904.

86. Mishra M, Chun D-M. α-Fe_2O_3 as a photocatalytic material: A review. *Applied Catalysis A: General* 2015; **498**: 126–41.

87. Liu M, Li H, Zeng Y. Facile Preparation of Efficient WO_3 Photocatalysts Based on Surface Modification. *Journal of Nanomaterials* 2015; **2015**: 1–7.

88. Vallejo W, Cantillo A, Díaz-Uribe C. Methylene Blue Photodegradation under Visible Irradiation on Ag-Doped ZnO Thin Films. *International Journal of Photoenergy* 2020; **2020**: 1–11.

89. Lavand AB, Malghe YS. Synthesis, characterization and visible light photocatalytic activity of carbon and iron modified ZnO. *Journal of King Saud University - Science* 2018; **30**: 65–74.

90. Yasmeen H, Zada A, Ali S, *et al.* Visible light-excited surface plasmon resonance charge transfer significantly improves the photocatalytic activities of ZnO semiconductor for pollutants degradation. *Journal of the Chinese Chemical Society* 2020; **67**: 1611–7.

91. Zhen Y, Yang C, Shen H, *et al.* Photocatalytic performance and mechanism insights of a S-scheme g-C_3N_4/Bi_2MoO_6 heterostructure in phenol degradation and hydrogen evolution reactions under visible light. *Phys Chem Chem Phys* 2020; **22**: 26278–88.

92. Reginato G, Zani L, Calamante M, Mordini A, Dessì A. Dye-Sensitized Heterogeneous Photocatalysts for Green Redox Reactions. *European Journal of Inorganic Chemistry* 2020; **2020**: 899–917.

93. Kumar SG, Devi LG. Review on modified TiO_2 photocatalysis under UV/visible light: selected results and related mechanisms on interfacial charge carrier transfer dynamics. *J Phys Chem A* 2011; **115**: 13211–41.
94. Wang S, Tian H, Ren C, Yu J, Sun M. Electronic and optical properties of heterostructures based on transition metal dichalcogenides and graphene-like zinc oxide. *Sci Rep* 2018; **8**: 12009.
95. Angel RD, Durán-Álvarez JC, Zanella R. TiO_2-Low Band Gap Semiconductor Heterostructures for Water Treatment Using Sunlight-Driven Photocatalysis. In: Dongfang Y, ed. *Titanium Dioxide - Material for a Sustainable Environment*, IntechOpen, London, UK, 2018.
96. Koswatta SO, Koester SJ, Haensch W. On the Possibility of Obtaining MOSFET-Like Performance and Sub-60-mV/dec Swing in 1-D Broken-Gap Tunnel Transistors. *IEEE Transactions on Electron Devices* 2010; **57**: 3222–30.
97. Rogé V, Didierjean J, Crêpellière J, *et al.* Tuneable Functionalization of Glass Fibre Membranes with ZnO/SnO_2 Heterostructures for Photocatalytic Water Treatment: Effect of SnO_2 Coverage Rate on the Photocatalytic Degradation of Organics. *Catalysts* 2020; **10**(7): 733.
98. Mahala C, Sharma MD, Basu M. Type-II Heterostructure of ZnO and Carbon Dots Demonstrates Enhanced Photoanodic Performance in Photoelectrochemical Water Splitting. *Inorganic Chemistry* 2020; **59**: 6988–99.
99. Enesca A, Andronic L. The Influence of Photoactive Heterostructures on the Photocatalytic Removal of Dyes and Pharmaceutical Active Compounds: A Mini-Review. *Nanomaterials* 2020; **10**(9): 1766.
100. Uddin MT, Hoque ME, Chandra Bhoumick M. Facile one-pot synthesis of heterostructure SnO_2/ZnO photocatalyst for enhanced photocatalytic degradation of organic dye. *RSC Advances* 2020; **10**: 23554–65.
101. Wang G, Geng L, Tang W, *et al.* Two dimensional CdS/ZnO type-II heterostructure used for photocatalytic water-splitting. *Nanotechnology* 2020; **31**: 485701.
102. Wang Y, Chen C, Tian W, Xu W, Li L. Designing $WO_3/CdIn_2S_4$ type-II heterojunction with both efficient light absorption and charge separation for enhanced photoelectrochemical water splitting. *Nanotechnology* 2019; **30**: 495402.
103. Chen H, Xie Y, Sun X, *et al.* Efficient charge separation based on type-II g-C_3N_4/ TiO_2-B nanowire/tube heterostructure photocatalysts. *Dalton Transactions* 2015; **44**: 13030–9.
104. Basu M, Garg N, Ganguli AK. A type-II semiconductor (ZnO/CuS heterostructure) for visible light photocatalysis. *J. Mater. Chem. A* 2014; **2**: 7517–25.
105. Nyamukamba P, Moloto MJ, Mungondori H. Visible Light-Active CdS/TiO_2 Hybrid Nanoparticles Immobilized on Polyacrylonitrile Membranes for the Photodegradation of Dyes in Water. *Journal of Nanotechnology* 2019; **2019**: 1–10.
106. Ng BJ, Putri LK, Kong XY, Teh YW, Pasbakhsh P, Chai SP. Z-Scheme Photocatalytic Systems for Solar Water Splitting. *Adv Sci (Weinh)* 2020; **7**: 1903171.
107. Bard AJ. Photoelectrochemistry and heterogeneous photo-catalysis at semiconductors. *Journal of Photochemistry* 1979; **10**: 59–75.
108. Low J, Jiang C, Cheng B, Wageh S, Al-Ghamdi AA, Yu J. A Review of Direct Z-Scheme Photocatalysts. *Small Methods* 2017; **1**: 1700080.
109. Pan Z, Zhang G, Wang X. Polymeric Carbon Nitride/Reduced Graphene Oxide/ Fe_2O_3: All-Solid-State Z-Scheme System for Photocatalytic Overall Water Splitting. *Angew Chem Int Ed Engl* 2019; **58**: 7102–6.
110. Osaki J, Yoda M, Takashima T, Irie H. Selective loading of platinum or silver cocatalyst onto a hydrogen-evolution photocatalyst in a silver-mediated all solid-state Z-scheme system for enhanced overall water splitting. *RSC Advances* 2019; **9**: 41913–7.

111. Ng B-J, Putri LK, Tan L-L, Pasbakhsh P, Chai S-P. All-solid-state Z-scheme photocatalyst with carbon nanotubes as an electron mediator for hydrogen evolution under simulated solar light. *Chemical Engineering Journal* 2017; **316**: 41–9.
112. Wang L, Li J, Wang Y, *et al.* Construction of 1D SnO_2-coated ZnO nanowire heterojunction for their improved n-butylamine sensing performances. *Sci Rep* 2016; **6**: 35079.
113. Huang Z-F, Song J, Wang X, *et al.* Switching charge transfer of $C_3N_4/W_{18}O_{49}$ from type-II to Z-scheme by interfacial band bending for highly efficient photocatalytic hydrogen evolution. *Nano Energy* 2017; **40**: 308–16.
114. Lin H, Zhao L. Novel g-C_3N_4/TiO_2 nanorods with enhanced photocatalytic activity for water treatment and H2 production. *Journal of Materials Science: Materials in Electronics* 2019; **30**: 18191–9.
115. Paul DR, Gautam S, Panchal P, Nehra SP, Choudhary P, Sharma A. ZnO-Modified g-C_3N_4: A Potential Photocatalyst for Environmental Application. *ACS Omega* 2020; **5**: 3828–38.
116. Jo W-K, Lee JY, Natarajan TS. Fabrication of hierarchically structured novel redox-mediator-free $ZnIn_2S_4$ marigold flower/Bi_2WO_6 flower-like direct Z-scheme nanocomposite photocatalysts with superior visible light photocatalytic efficiency. *Physical chemistry chemical physics : PCCP* 2016; **18**: 1000–16.
117. Wang S, Zhu B, Liu M, Zhang L, Yu J, Zhou M. Direct Z-scheme ZnO/CdS hierarchical photocatalyst for enhanced photocatalytic H_2-production activity. *Applied Catalysis B: Environmental* 2019; **243**: 19–26.
118. Venkatachalam N, Palanichamy M, Arabindoo B, Murugesan V. Enhanced photocatalytic degradation of 4-chlorophenol by Zr^{4+} doped nano TiO_2. *Journal of Molecular Catalysis A: Chemical* 2007; **266**: 158–65.
119. Roushenas P, Ong ZC, Ismail Z, *et al.* Operational parameters effects on photocatalytic reactors of wastewater pollutant: a review. *Desalination and Water Treatment* 2018; **120**: 109–18.
120. Sun M, Guo P, Wang M, Ren F. The effect of pH on the photocatalytic performance of $BiVO_4$ for phenol mine sewage degradation under visible light. *Optik* 2019; **179**: 672–9.
121. Guo M, Wang L, Cai Y, *et al.* Effect of pH Value on Photocatalytic Performance and Structure of $AgBr/Ag_2CO_3$ Heterojunctions Synthesized by an In Situ Growth Method. *Journal of Electronic Materials* 2020; **49**: 3301–8.
122. Intaphong P, Phuruangrat A, Karthik K, Dumrongrojthanath P, Thongtem T, Thongtem S. Effect of pH on Phase, Morphology and Photocatalytic Properties of BiOBr Synthesized by Hydrothermal Method. *Journal of Inorganic and Organometallic Polymers and Materials* 2020; **30**: 714–21.
123. Herrmann J-M. Fundamentals and misconceptions in photocatalysis. *Journal of Photochemistry and Photobiology A: Chemistry* 2010; **216**: 85–93.
124. Ariza-Tarazona MC, Villarreal-Chiu JF, Hernández-López JM, *et al.* Microplastic pollution reduction by a carbon and nitrogen-doped TiO_2: Effect of pH and temperature in the photocatalytic degradation process. *Journal of Hazardous Materials* 2020; **395**: 122632.
125. Andrade Neto NF, Oliveira PM, Bomio MRD, Motta FV. Effect of temperature on the morphology and optical properties of Ag_2WO_4 obtained by the co-precipitation method: Photocatalytic activity. *Ceramics International* 2019; **45**: 15205–12.
126. Wang Y, Hong C-S. Effect of hydrogen peroxide, periodate and persulfate on photocatalysis of 2-chlorobiphenyl in aqueous TiO_2 suspensions. *Water Research* 1999; **33**: 2031–6.

127. Boarini P, Carassiti V, Maldotti A, Amadelli R. Photocatalytic Oxygenation of Cyclohexane on Titanium Dioxide Suspensions: Effect of the Solvent and of Oxygen. *Langmuir* 1998; **14**: 2080–5.
128. Vinu R, Madras G. Photocatalytic Degradation of Water Pollutants Using Nano-TiO_2. In: Zang L, ed. *Energy Efficiency and Renewable Energy Through Nanotechnology. Green Energy and Technology*: Springer, London, 2011: 625–77.
129. Boruah PK, Borthakur P, Darabdhara G, *et al.* Sunlight assisted degradation of dye molecules and reduction of toxic Cr(vi) in aqueous medium using magnetically recoverable Fe_3O_4/reduced graphene oxide nanocomposite. *RSC Advances* 2016; **6**: 11049–63.
130. Anju Chanu L, Joychandra Singh W, Jugeshwar Singh K, Nomita Devi K. Effect of operational parameters on the photocatalytic degradation of Methylene blue dye solution using manganese doped ZnO nanoparticles. *Results in Physics* 2019; **12**: 1230–7.
131. Fosso-Kankeu E, Pandey S, Sinha Ray S. *Photocatalysts in Advanced Oxidation Processes for Wastewater Treatment*: Scrivener Publishing LLC, Beverly, Massachusetts, United States, 2020.
132. Sohrabnezhad S, Pourahmad A, Radaee E. Photocatalytic degradation of basic blue 9 by CoS nanoparticles supported on AlMCM-41 material as a catalyst. *J Hazard Mater* 2009; **170**: 184–90.
133. Paul DR, Sharma R, Nehra SP, Sharma A. Effect of calcination temperature, pH and catalyst loading on photodegradation efficiency of urea derived graphitic carbon nitride towards methylene blue dye solution. *RSC Advances* 2019; **9**: 15381–91.
134. Terzian R, Serpone N. Heterogeneous photocatalyzed oxidation of creosote components: mineralization of xylenols by illuminated TiO_2 in oxygenated aqueous media. *Journal of Photochemistry and Photobiology A: Chemistry* 1995; **89**: 163–75.
135. Chen Q, Chen L, Qi J, *et al.* Photocatalytic degradation of amoxicillin by carbon quantum dots modified $K_2Ti_6O_{13}$ nanotubes: Effect of light wavelength. *Chinese Chemical Letters* 2019; **30**: 1214–8.
136. Aghdam TR, Mehrizadeh H, Salari D, Tseng H-H, Niaei A, Amini A. Photocatalytic removal of NOx over immobilized $BiFeO_3$ nanoparticles and effect of operational parameters. *Korean Journal of Chemical Engineering* 2018; **35**: 994–9.

doi: 10.2166/9781789061932_0037

Chapter 2
Metal organic frameworks for photocatalytic water treatment

Mohammadreza Beydaghdari[1], Mehrdad Asgari[2,3] and Fahimeh Hooriabad Saboor[1]*

[1]Department of Chemical Engineering, University of Mohaghegh Ardabili, Ardabil, Iran
[2]Department of Chemical Engineering & Biotechnology, University of Cambridge, Philippa Fawcett Drive, Cambridge CB3 0AS, UK
[3]Institute of Chemical Sciences and Engineering, École Polytechnique, Fédérale de Lausanne (EPFL), EPFL-ISIC-Valais, Sion 1950, Switzerland
*Corresponding author: f.saboor@uma.ac.ir

ABSTRACT

This chapter focuses on recent advances in the application of photocatalytic metal-organic frameworks (MOFs) for water treatment. Degradation mechanism and photocatalytic performance of MOFs for the removal of various pollutants in aquatic environments, such as dyes, pharmaceuticals, personal care products, heavy metals, pesticides, herbicides, and other organic compounds, are reviewed. Unique features of MOFs such as high porosity, tunable structural properties, and facile and reliable synthesis procedures, provide enthusiasm for the development of novel MOF-based composites and functionalized MOF structures that are used to enhance the photocatalytic removal of pollutants in water. This review aims to highlight the essential features and critical limits of MOF-based photocatalytic adsorbents under visible, solar, or ultraviolet (UV) light irradiation towards the ultimate goal of having more efficient large-scale water treatment processes.

2.1 INTRODUCTION

With the development and rapid evolution of industries, applying novel and modern chemical transformation routes seems necessary. They are especially important considering that such industrial growth has led to an increased amount of waste production. Daily activities of humans, processing of chemicals in industries, and agriculture leads to many contaminants being released into the water resources [1]. More than half of the world's population lives in areas with limited access to clean water. Therefore, providing a clean water supply to all people globally is a significant challenge, with population growth and global warming exacerbating

© 2022 The Editors. This is an Open Access book chapter distributed under the terms of the Creative Commons Attribution Licence (CC BY-NC-ND 4.0), which permits copying and redistribution for noncommercial purposes with no derivatives, provided the original work is properly cited (https://creativecommons.org/licenses/by-nc-nd/4.0/). This does not affect the rights licensed or assigned from any third party in this book. The chapter is from the book Photocatalytic Water and Wastewater Treatment, Alireza Barzagan (Ed.).

the situation [2]. As water is a crucial element for the continuance of life on Earth [3, 4], the global population coupled with the scarcity of water resources is considered one of the most critical problems of today's world [5]. Polluted water is a severe threat to the life of humans, plants, aquatic organisms, and other living things. Therefore, it is vital to remove the available pollutants from industrial effluents before discharging them to the environment.

Currently, several methods exist to remove contaminants, based on different methods such as filtration, coagulation, membrane separation, adsorption, and biological processes [6–9]. These methods are not ideal due to low efficiency, high cost, and the need for complex instruments [10]. An alternative to conventional water treatment methods is photocatalytic degradation, an efficient wastewater treatment which does not create secondary pollution. Various photocatalysts including metal oxides (e.g., TiO_2, NiO, WO_3, and ZnO) [11, 12], halides (e.g., BiOI and AgCl) [13], metal sulfides (e.g., CdS, In_2S_3, ZnS and MoS_2) [14, 15], and non-metal semiconductors (e.g., $g-C_3N_4$ and graphene) are used for the photocatalytic removal of contaminants from aqueous solutions. Metal-organic frameworks (MOFs) are a new class of porous materials with a large surface area and high chemical tunability, used in a wide range of applications, including sensors, biomedicine, catalysis, gas separation, super-capacitors, and adsorption [16]. The wide applications of MOFs are mainly due to their structural features such as their porous nature, reusability, low framework density, and high stability [17–21]. MOFs have shown enhanced performance in photocatalytic applications, particularly the degradation of organic pollutants. The large surface area and high chemical tunability of MOFs are two important properties that make them suitable candidates for photocatalytic removal of organic pollutants in aqueous solutions [22]. Accordingly, several different frameworks, such as NTU-9, MIL-53, MIL-125, MIL-88, MIL-101, UiO-66, and MOF-5 are used in the photocatalytic degradation of pollutants.

2.2 BASIS OF PHOTOCATALYTIC DEGRADATION

Photocatalysis is a clean and environmentally-friendly phenomenon for removing many organic pollutants by converting them into less hazardous compounds, without any secondary pollution. Photocatalytic degradation is typically carried out through an advanced oxidative process in the presence of a photocatalyst [23]. Photocatalytic degradation of pollutants is considered one of the most promising solutions for the removal of organic pollutants due to its low technical complexity, high stability, good efficiency, low operating costs, mineralization of the intermediates without leaving secondary pollution, and ambient operating conditions [24] compared to conventional methods [25, 26]. Therefore, photocatalysts have been utilized in various applications such as water and air purification systems, hydrogen evolution, CO_2 reduction, and water splitting [27, 28].

The chemical conversion mechanism through photocatalysis includes photocatalytic conversion of reactants by absorbing light and creating electron-hole pairs. The reactants then regenerate their chemical composition after each cycle [29]. Accelerating oxidation and reduction reactions via solar energy is the

Metal organic frameworks for photocatalytic water treatment

main feature of a photocatalytic material. Photocatalysts have two separate energy bands: the lowest energy band where no electrons exist, called the conduction band (CB), and the highest energy band where there are some electrons, which is called the valence band (VB). The energetic barrier between the CB and VB is known as the bandgap energy (Eg) [30]. By light radiation, the energy of which is higher than or equal to the bandgap energy of the photocatalyst, electrons (e$^-$) are excited to the CB due to receiving this energy. Therefore, positive charge carriers, that is, holes (h$^+$), are formed in the VB [31]. The excited energy can be released as heat by photogenerated charges. Also, photogenerated charges can transfer to the photocatalyst surface, generate reactive oxygen species (ROS), or recombine immediately [32]. The excited electrons can reduce adsorbed oxygen molecules on the CB, and the positive holes oxidize water to produce hydroxyl radicals, as shown in Figure 2.1. The ·OH free radicals then attack organic groups of the pollutant and convert them into some non-toxic organic species or completely degrade them into CO_2 and H_2O [33]. During the photocatalytic process, the adsorbed surface contaminants, such as dyes and pharmaceuticals, react with the free electrons/holes and oxidizing species, resulting in the degradation of contaminants. The mechanism of activation of a photocatalyst by ultraviolet (UV) light irradiation, which is carried out through a few consecutive steps, can be seen in the following:

$$\text{photocatalyst} + h\nu \rightarrow e^- + h^+ \quad (2.1)$$

$$e^- + O_2 \rightarrow O_2^{\cdot -} \quad (2.2)$$

$$h^+ + \text{Organic}(R) \rightarrow \text{Intermediates} \rightarrow CO_2 + H_2O \quad (2.3)$$

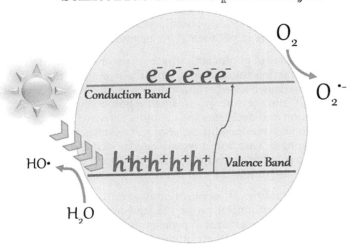

Figure 2.1 Photocatalysis mechanism in semiconductor materials [34].

$$h^+ + H_2O \rightarrow \ ^\cdot OH + H^+ \tag{2.4}$$

$$^\cdot OH + Organic\,(R) \rightarrow Intermediates \rightarrow CO_2 + H_2O \tag{2.5}$$

Among various factors affecting the photocatalytic degradation efficiency, the most critical factor is the photocatalyst's ability to produce long-lived electrons and holes. Other factors include light source, light intensity, organic pollutants nature, pH, temperature, and solvent [35, 36]. A wide range of photocatalysts have been used for photocatalytic degradation of organic compounds, such as metal oxides (e.g., TiO_2, ZnO, WO_3, CeO_2, Fe_2O_3, and Bi_2O_3) [37, 38], metal sulfides (e.g., ZnS, CdS, MoS_2, CuS, and Bi_2S_3) [12, 39], non-metal semiconductors (e.g., graphene and g-C_3N_4) [40, 41], and multicomponent materials (e.g., $Bi_2O_2CO_3$/ Bi_2O_4 and Bi_2S_3/Bi_2O_3/$Bi_2O_2CO_3$) [42, 43].

Oxide-based heterogeneous semiconductors are promising candidates for photocatalytic applications due to their enhanced light absorbing properties, lifetime of the excited state, variable surface chemistry, electronic structure, and charge transfer properties [44, 45]. Among these materials, TiO_2, ZnS, CdS, and ZnO exhibit effective pollutant degradation and complete mineralization in environmental applications [46]. TiO_2 is a suitable catalyst for degrading and removing organic pollutants because of its non-toxicity and long-term stability [47]. It can be used as a photocatalyst in various applications such as water treatment, air purification, and surface self-cleaning. TiO_2 has three crystal structures, including anatase, rutile, and brookite [48, 49]. Among these crystal structures, anatase has the best photocatalytic activity due to its lower recombination rate and longer carrier lifetime [50, 51].

2.3 METAL-ORGANIC FRAMEWORKS AS A PHOTOCATALYST

Despite the advantages mentioned above, there are some limitations associated with the structural properties of semiconductor-type photocatalysts, which have limited their usage in photocatalytic degradation:

(1) Low efficiency of photocatalyst recovery from the reaction environment after completing the reaction, which is necessary to purify water, is a significant drawback of semiconductor photocatalysts such as TiO_2.

(2) The high surface energy of small photocatalyst particles promotes aggregation and decreases the catalytic activity [52].

(3) Conventional semiconductor photocatalysts have low-stability in the aqueous phase. Therefore, they can be corroded and dissolved as the metal ion in the solution under light irradiation. Such corrosion/degradation is due to the migration of metal ions into the aqueous solution [53, 54].

Due to the limitations of conventional semiconductor materials, novel photocatalysts with enhanced properties and higher efficiency are required. Metal-organic frameworks are a promising alternative for photocatalytic removal of pollutants due to their unique properties, including unprecedently high surface area, high porosity, very high crystallinity, high stability, flexible synthesis strategies, reusability, high density of active sites, and

Metal organic frameworks for photocatalytic water treatment 41

three-dimensional (3D) framework structure [55–57]. MOFs are used in many different applications such as catalysis [58], drug delivery [59], gas storage [60], and sensing [61]. MOFs are composed of two main building blocks: organic linkers and metal ion nodes or clusters [62]. According to the hard/soft/acid/base (HSAB) principle and metal-ligand bond strengths, stable MOFs can be synthesized using either a soft or hard Lewis base, depending on the metal site's choice [63]. Divalent metal ions such as Ni^{2+}, Cu^{2+}, Zn^{2+}, and Co^{2+} can form stable MOFs with soft Lewis bases, while high-valent metal ions such as Fe^{3+}, Ti^{4+}, Al^{3+}, Fe^{3+}, Cr^{3+}, and Zr^{4+} make a more robust framework through coordination with hard Lewis bases. Some stable MOFs include the MIL series (Material Institute Lavoisier) and ZIFs (zeolitic imidazolate frameworks) [64, 65]. Some of the unique features of MOFs for catalytic removal of organic pollutants from aqueous solutions, compared to those of conventional semiconductors, are as follows [66–70]:

(i) Easier diffusion of target molecules in MOF structure due to their high porosity.
(ii) More efficient use of solar energy due to the existence of tunable active sites.
(iii) The presence of a single active metal site inside MOF structure due to their unique structure.
(iv) Rational reticular design of MOF structure at the molecular level due to their high modularity and tunability.
(v) High crystallinity of MOFs provides a full characterization of their structure, which helps to perform detailed studies on the photocatalytic process.
(vi) Controlling the topology of MOF structure, which mainly originates from the presence and possibility of using various organic ligands, enables the formation of different pore shapes and sizes, each of which could be optimal for a particular application.

These unique properties of MOFs allow them to resolve the shortcomings of the conventional photocatalysts [71, 72]. The first use of MOFs to remove organic pollutants was reported in 2007 by Garcia and coworkers, in which MOF-5 was used to degrade phenol [73, 74]. In recent years, many catalytic studies have been performed on removing and degrading organic pollutants using MOFs, mainly due to their reusability and ease of use [75, 76]. Under light irradiation, organic ligands in MOFs can absorb light and activate metal sites through the ligand-to-metal charge transfer mechanism (LMCT). In this process, photoinduced electrons are generated [77], which can be transferred from the highest occupied molecular orbital (HOMO) to the lowest unoccupied molecular orbital (LUMO) of the MOFs. Consequently, these charge carriers are transferred to the surface of metal-oxo [78]. The role of LUMO/HOMO in MOFs is similar to the VB/CB in semiconductors. The superoxide radicals are formed by transferring the photogenerated electrons of LUMO orbitals to the adsorbed O_2 molecules. Meanwhile, surface hydroxyl group/water are oxidized in HOMO holes, and hydroxyl radicals are generated [34], as shown in Figure 2.2.

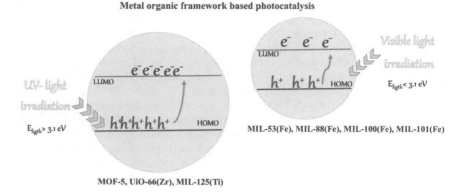

Figure 2.2 Photocatalysis mechanism of MOFs [34].

In addition to the LMCT mechanism, metal-to-metal charge transfer and ligand-to-ligand charge transfer have also been used to explain photocatalytic processes in MOFs [79]. Several factors such as the type of pollutants, energy band, and structure of MOFs can affect the photocatalytic mechanism [80]. Despite the decent performance of MOFs in photocatalytic degradation, several studies target more efficient charge generation, extended visible light absorption, and high stability/recyclability to further enhance their efficiency. The use of ligands that absorb a wide range of visible light is one of the suitable and efficient strategies for developing MOFs with decent light absorption in the visible region. Chemical functional groups such as -NH_2, -SH and NO_2 and some metal complexes with high absorption coefficients in the visible region are among those molecular species which could be appended inside MOF structures to enhance visible light absorption [81]. These functional groups, especially the amine groups, can enhance the electron transfer from the excited functionalized organic linker to oxo-metal clusters. Also, MOF structures with large secondary building units have semiconducting behavior with smaller bandgap energies and can absorb visible light more efficiently [82].

Metal ion doping into the inorganic metal clusters of MOF structures can also enhance the photocatalytic performance, similar to inorganic semiconductor photocatalysts [83, 84]. Due to the various structural features of MOFs, it is possible to design and synthesize different bimetallic MOF-based photocatalysts with enhanced photocatalytic performance. For example, NH_2-UiO-66, which consists of Zr/Ti-based clusters, has shown an enhanced photocatalytic activity for different photocatalytic applications, namely H_2 evolution and CO_2 reduction reactions [85].

The introduction of nanoparticles (NPs) can enhance the performance of MOF-based photocatalysts. Under light irradiation, metal NPs can act as an electron reservoir from the MOF excitation band and, at the same time, can improve the separation efficiency. Due to tunable cavity dimensions and high porosity, MOFs show more advantages in embedding NPs than other

Metal organic frameworks for photocatalytic water treatment 43

conventional semiconductors. Since some noble metals such as Au, Pt, and Pd have shown high efficiency for degradation of the organic pollutants, embedding such NPs in the MOF structure could yield highly efficient photocatalysts. To maximize the synergetic effects between the host structure and the embedded NPs, an appropriate selection of metal NPs and MOFs is crucial [86]. To date, the synthesis of multifunctional MOF-based nanocomposites for photocatalytic applications has received much attention. In addition to the noble metal NPs, the incorporation of semiconductor NPs such as TiO_2, ZnO, $BiVO_4$, Bi_2WO_6, and graphene can also tune the optical properties of the host MOF, thereby improving the photocatalytic performance.

Many organic pollutants present in the environment, such as dyes, pesticides, pharmaceuticals and personal care products (PPCPs), and herbicides, can have harmful effects on humans and other living species [87, 88]. Photocatalytic degradation using MOFs and MOF-based catalysts can be a suitable way to efficiently degrade and remove pollutants from the environment [89].

2.3.1 Photocatalytic degradation of dyes using MOFs

Dyes are among the most critical and dangerous pollutants in various industries such as textile, food processing, leather tanning, rubber and plastic, printing, and so on. [90–92]. Dyes can be synthetic or natural (e.g., vegetable, mineral sources, or animals), mostly contain a complex organic molecule, and are generally composed of two main components, including auxochromes and chromophores. Auxochromes have functional groups such as $-NH_2$, NHR, -OH, NR_2, and -COOH, which are electron donors. These chemical functionalities increase the solubility of dyes in aqueous solutions and enhance the dye adhesion to fibers. The chromophores contain heteroatoms, such as O, S, and N, with non-bonding electrons. These functionalities are electron receivers and are responsible for dye formation [93–95]. Dyes are stable under light exposure, heat, and elevated temperature. Such long-term stability, in addition to their toxicity, makes dyes stable pollutants in the water resources, endangering human health, aquatics, and plants [96]. Therefore, removing dyes from industrial effluents and other water sources is one of the main priorities in achieving access to clean water.

Among the different methods applied, photocatalytic degradation using MOFs and MOF-based materials is an effective method for removing dyes from wastewater. In photocatalytic degradation, the dyes are decomposed into small molecules such as CO_2 and H_2O through the generation of hydroxyl radicals. An efficient photocatalyst should have unique properties, namely optimum CB and VB energy levels, fast electron-hole transport, efficient charge separation, stability in an aqueous solution, and sufficient light absorption capacity [97]. Dyes undergo color change upon degradation, making it possible to follow the degradation visually and characterize it by UV-Vis spectrophotometry. Some recent results on dye degradation using MOF-based photocatalysts are summarized in Table 2.1.

Lou-Hong Zhang *et al.* [98] synthesized a 3D copper organic framework of HKUST-1 (MOF-199) using a solvothermal method and applied it for the degradation of RB13 dye. This framework shows a degradation efficiency

Table 2.1 Some recent studies on photocatalytic degradation of dyes using MOFs.

MOF	Band Gap (eV)	Dye Target[a]	Molecular Formula	Degradation Efficiency (%)	Light Source[a]	Time (min)	References
HKUST–1	3.5	RB13	$C_{29}H_{16}ClN_7Na_4O_{14}S_4$	70	Vis	–	[98]
ZIF–8	5.16	MB	$C_{16}H_{18}ClN_3S$	82.3	UV	120	[99]
$Zn_2(odpt)(bpy)(H_2O)(bpy)_{0.5}$	3.6	MO	$C_{14}H_{14}N_3NaO_3S$	91.7	UV	120	[100]
$Zn_2(odpt)(bpy)(H_2O)(bpy)_{0.5}$	3.6	MB	$C_{16}H_{18}ClN_3S$	90.9	UV	120	[100]
MOF–1	3.87	MB	$C_{16}H_{18}ClN_3S$	88	UV	80	[101]
MOF–2	2.30	MB	$C_{16}H_{18}ClN_3S$	91	UV	80	[101]
MOF–3	1.99	MB	$C_{16}H_{18}ClN_3S$	92	UV	80	[101]
MOF–1	3.87	MO	$C_{14}H_{14}N_3NaO_3S$	72	UV	120	[101]
MOF–2	2.30	MO	$C_{14}H_{14}N_3NaO_3S$	76	UV	120	[101]
MOF–3	1.99	MO	$C_{14}H_{14}N_3NaO_3S$	81	UV	120	[101]
MOF–199	5.43	BB41	$C_{20}H_{26}N_4O_6S_2$	99	UV	180	[102]
TMU–4	2.6	CR	$C_{32}H_{22}N_6Na_2O_6S_2$	93.8	UV	120	[103]
TMU–5	2.1	CR	$C_{32}H_{22}N_6Na_2O_6S_2$	87.8	UV	120	[103]
TMU–6	2.2	CR	$C_{32}H_{22}N_6Na_2O_6S_2$	98.3	UV	90	[103]
UiO–66(Ti)	3.2	MB	$C_{16}H_{18}ClN_3S$	87.1	UV-Vis	160	[104]
MIL–53(Fe)	2.72	MB	$C_{16}H_{18}ClN_3S$	99	UV-Vis	20	[105]
MIL–88A	2.05	MB	$C_{16}H_{18}ClN_3S$	100	Vis	20	[106]
MIL–100(Fe)	3.08	MO	$C_{14}H_{14}N_3NaO_3S$	64	UV	60	[107]
NTU–9	1.72	RhB	$C_{28}H_{31}ClN_2O_3$	100	Vis	20	[108]
NTU–9	1.72	MB	$C_{16}H_{18}ClN_3S$	100	Vis	20	[108]
$[Cd(4,4\text{-}bpy)(H_2O)(S_2O_3)].2H_2O$	2.91	MB	$C_{16}H_{18}ClN_3S$	70	Sunlight	90	[109]
$Cd(tdc)(bix)(H_2O)$	3.31	MO	$C_{14}H_{14}N_3NaO_3S$	90	UV	150	[110]
Cu/ZIF–67	1.95	MO	$C_{14}H_{14}N_3NaO_3S$	100	Vis	25	[111]
$Zn_2(4,4\text{-}bpy)](4,4\text{-}obb)_2$	4.02	RBBR	$C_{22}H_{16}N_2Na_2O_{11}S_3$	80	UV	100	[112]
$Ni_2(4,4\text{-}bpy](4,4\text{-}obb)_2.2H_2O$	3.89	RBBR	$C_{22}H_{16}N_2Na_2O_{11}S_3$	95	UV	100	[112]

(Continued)

Table 2.1 (Continued) Some recent studies on photocatalytic degradation of dyes using MOFS.

MOF	Band Gap (eV)	Dye Target[a]	Molecular Formula	Degradation Efficiency (%)	Light Source[a]	Time (min)	References
MIL–101	3.74	RB	$C_{26}H_{21}N_5Na_4O_{19}S6$	100	UV	45	[113]
Ca/TiO$_2$/NH$_2$–MIL–125	3.2	MO	$C_{14}H_{14}N_3NaO_3S$	87.29	Vis	120	[114]
Ca/TiO$_2$/NH$_2$–MIL–125	3.2	RhB	$C_{28}H_{31}ClN_2O_3$	83.12	Vis	120	[114]
NH$_2$–MIL–125(Ti)(NMT)	2.89	RhB	$C_{28}H_{31}ClN_2O_3$	89	Vis	30	[115]
Pd/rGO/MIL–101(Cr)	–	Brilliant green	$C_{27}H_{33}N_2HO_4S$	100	Vis	15	[116]
Pd/rGO/MIL–101(Cr)	–	Acid fuchsin	$C_{20}H_{17}N_3Na_2O_9S_3$	100	Vis	15	[116]
GO/MIL–101(Cr)	2.17	MG	$C_{23}H_{25}ClN_2$	92	Sunlight	60	[117]
MOF–1/GO/Fe$_3$O$_4$	2.39	MB	$C_{16}H_{18}ClN_3S$	95	Vis	80	[118]
rGO/NH$_2$–MIL–125(Ti)	–	MB	$C_{16}H_{18}ClN_3S$	100	Vis	30	[119]
Bi$_2$WO$_6$/UiO–66	–	RhB	$C_{28}H_{31}ClN_2O_3$	100	Vis	180	[120]
BiVO$_4$/UiO–66	–	RhB	$C_{28}H_{31}ClN_2O_3$	100	Vis	140	[121]
Bi$_2$MoO6/UiO–66	–	RhB	$C_{28}H_{31}ClN_2O_3$	95	Vis	120	[122]
Bi$_2$MoO$_6$/MIL–100	–	RhB	$C_{28}H_{31}ClN_2O_3$	90	Vis	90	[123]
BiOBr/NH$_2$–MIL–125	–	RhB	$C_{28}H_{31}ClN_2O_3$	100	Vis	100	[124]
AgI-MIL–53(Fe)	2.29	RhB	$C_{28}H_{31}ClN_2O_3$	100	Vis	45	[125]
AgI-MIL–53(Fe)	2.29	MO	$C_{14}H_{14}N_3NaO_3S$	65	Vis	45	[125]
AgI-MIL–53(Fe)	2.29	AB	$C_{20}H_{13}N_2NaO_5S$	39	Vis	45	[125]
Bi$_2$MoO$_6$/UiO–66	–	RhB	$C_{28}H_{31}ClN_2O_3$	96	Vis	60	[122]
Ag/Ag$_3$PO$_4$/HKUST–1	–	PBS	$C_{22}H_{14}N_4Na_2O_7S_2$	80	Vis	90	[126]
Ag$_3$PO$_4$/MIL–101/NiFe$_2$O$_4$	–	RhB	$C_{28}H_{31}ClN_2O_3$	95	Vis	30	[127]
C$_3$N$_4$/MIL–125(Ti)	–	RhB	$C_{28}H_{31}ClN_2O_3$	95.2	Vis	30	[128]
UiO–66/g-C$_3$N$_4$	–	MB	$C_{16}H_{18}ClN_3S$	100	Vis	240	[129]
MIL–53(Fe)	2.88	RhB	$C_{28}H_{31}ClN_2O_3$	100	Vis(H$_2$O$_2$)	50	[130]
CuO/ZIF–8	4.93	Rh6G	$C_{28}H_{31}N_2O_3Cl$	99.5	Sunlight	45	[131]
MIL–53(Fe)	2.81	AO7	$C_{16}H_{11}N_2NaO_4S$	100	Vis	90	[132]
NiCo-LDHs	–	RR–120	$C_{44}H_{24}C_{12}N_{14}N_6O_{20}S_6$	89	Vis(PMS)	10	[133]

[a]The abbreviations are as follows: RB13, reactive blue 13; MB, methylene blue; MO, methyl orange; BB41, basic blue 41; CR, congo red; RhB, rhodamine B; RBBR, Remazol brilliant blue R; RB, Remazol black; MG, malachite green; AB, alkali blue; PBS, phosphate-buffered saline; Rh6G, rhodamine 6G; AO7, acid orange 7; RR-120, reactive red 120; PMS, peroxymonosulfate.

of 74% for RB13. Jiang *et al.* [99] reported the application of ZIF-8 for photocatalytic removal of methylene blue (MB) under UV light irradiation. The framework could effectively degrade 82.3% of methylene blue in just 120 min. Furthermore, they determined that ZIF-8 has a high absorption capacity in alkali media and shows good performance over a wide pH range. A new 3D and highly stable Zn(II)-based MOF, $[Zn_2(odpt)(bpy)(H_2O)(bpy)_{0.5}]$, was synthesized by Wang *et al.* [100]. Zn(II)-based MOF showed excellent performance in photocatalytic removal of methyl orange (MO) under UV light irradiation. The conversion of MO was reported to be 91.7% within 2 h after starting the reaction. Mahmoodi *et al.* [102] synthesized a Cu-based MOF, MOF-199, using a solvothermal method for degradation of 99% of basic blue 41 under UV light irradiation. Wang *et al.* [104] synthesized UiO-66 (Ti) with photoactive Ti sites, using the facial modified post-grafting method. Introduction of Ti into the MOF structure enhanced the optical properties of UiO-66 due to the formation of oxo-bridged hetero-Zr-Ti clusters and showed a removal efficiency of 87.1% for MB. Ahmadpour *et al.* [114] reported the synthesis of a $Ca/TiO_2/NH_2$-MIL-125 framework for the photodegradation of rhodamine B (RhB) and MO dyes. The $Ca/TiO_2/NH_2$-MIL-125 structure exhibited high stability, reusability, and high efficiency for removing MO and RhB with, respectively, 87.29% and 83.12% removal efficiency under visible-light irradiation.

A promising candidate of MOFs as a photocatalyst is MIL-type MOFs (Material of Institute Lavoisier), which was first reported by Jing-Jing Du *et al.* [105] (MIL-53(Fe)) for decolorization of MB dye in 2011. Since then, many types of research have been and continue to be conducted on MIL-type frameworks. MIL-101 was synthesized by Du *et al.* [113] to study the photocatalytic degradation of Remazol black B (RBB). This framework showed high surface area, high stability in the aqueous phase, and high crystallinity. The RBB was entirely removed in 45 min under UV irradiation. After four cycles, the photocatalytic efficiency showed a decrease of only 5%.

The development of visible-light-responsive MOFs as photocatalysts has attracted much attention in recent years. Xu *et al.* [106] synthesized hexagonal microrods of MIL-88A with the advantages of low cost and visible light response. The framework showed an operational performance for MB dye decolorization under visible-light irradiation with a degradation efficiency of 100% in 20 min and first-order kinetics of degradation. For degradation of RhB dye, Mahmoodi *et al.* [115] prepared an NH_2-MIL-125(Ti) composite using a hydrothermal method and successfully magnetized it with $CoFe_2O_4$ nanoparticles. A plasmonic photocatalyst, that is, Ag/AgCl, was assembled on NH_2-MIL-125(Fe) by an in situ deposition method to enhance the photocatalytic performance of RhB. Also, the photocatalytic efficiency was maintained after seven recycling times, showing high reusability of the framework. Wu *et al.* [116] synthesized Pd/GO/MIL-101(Cr) for relatively rapid photocatalytic degradation of two triphenylmethane dyes (acid fuchsin and brilliant green) in 15 min under visible-light irradiation. Gao *et al.* [108] reported on a Ti-based MOF, that is, NTU-9, for complete degradation of RhB and MB dyes. NTU-9, with a decent photocatalytic performance acts as a p-type semiconductor and is photoactive under visible-light irradiation.

Metal organic frameworks for photocatalytic water treatment 47

Combining MOF with nanostructured materials can improve the photocatalytic efficiency of MOF materials. Fazli et $al.$ [117] synthesized GO/MIL-101(Cr) and used it for photocatalytic degradation of malachite green (MG) dye. Under sunlight irradiation, the degradation efficiency was measured to be 92% after 1 h. A recoverable and stable photocatalyst, MOF-1/GO/Fe$_3$O$_4$, was reported on by Bai et $al.$ [118] and showed a degradation efficiency of 95% for MB. The excitation wavelength of MOF-1/GO/Fe$_3$O$_4$ exhibited a shift from the UV to the visible region. The removal efficiency of MOF-1/GO/Fe$_3$O$_4$ was improved by about 90%, compared to MOF-1. GO addition minimized the recombination rate of electron-hole pairs and accelerated the electron migration of the composite photocatalyst. A highly porous hybrid nanocomposite, reduced GO/NH$_2$-MIL-125(Ti), was synthesized by Huang et $al.$ [119] as a photocatalyst for MB removal. The hybrid photocatalyst entirely removed MB under visible-light irradiation in 30 min.

Another effective way to achieve highly efficient MOF-based photocatalysts is to combine MOFs with light-harvesting semiconductor materials such as Bi$_2$WO$_6$, BiOBr, Ag$_2$CO$_3$, and AgI. Sha et $al.$ [120] synthesized UiO-66 and applied it for the removal of RhB dye. Incorporation of UiO-66 with bismuth tungstate could effectively enhance the visible-light degradation and resulted in an entire removal of RhB in only 3 h. RhB was completely mineralized under visible light in 45 min using AgI-MIL-53(Fe) composite synthesized by Han et $al.$ [125]. Also, the photodegradation of MO and acid blue (AB) using AgI-MIL-53(Fe) composite were 65% and 39%, respectively. Adding AgI to the structure of MOF could prevent recombination of photogenerated electron-hole pairs and improve the photocatalytic performance of MIL-53(Fe). The photodegradation activity of AgI-MIL-53(Fe) composite was at 70% of initial activity after five cycles. Bi$_2$MoO$_6$/UiO-66 composite with a Zr:Bi molar ratio of 1:2 was applied for visible-light degradation of RhB by Ding et $al.$ [122] with decent stability, recyclability, and high removal efficiency of 96%, compared to the bare UiO-66 and Bi$_2$MoO$_6$. Such enhancement could be due to an increase in solar energy absorption and a reduction in the recombination rate of charge carriers.

g-C$_3$N$_4$ is a narrow bandgap photocatalyst (2.7 eV) for the degradation of organic pollutants [134]. However, the photocatalytic activity of g-C$_3$N$_4$ is limited due to the fast recombination of photogenerated carriers, low surface area, and confined active sites [128]. To address this limitation, Wang et $al.$ [135] successfully synthesized a g-C$_3$N$_4$/MIL-125(Ti) composite using a solvothermal method. g-C$_3$N$_4$/MIL-125(Ti) showed special features including, a mesoporous structure, good thermal stability, enhanced visible-light absorption capacity, and large surface area. This hybrid material was more efficient and showed a 24 times higher rate for photodegradation of RhB compared to pure MIL-125(Ti) and g-C$_3$N$_4$. The photodegradation mechanism of RhB over g-C$_3$N$_4$/MIL-125(Ti) with a conversion of 95.2% is presented in the following:

$$RhB + hv \rightarrow RhB^+ \rightarrow RhB^* \rightarrow RhB^{+\cdot} + e^- \qquad (2.6)$$

$$g - C_3N_4 + hv \rightarrow g - C_3N_4(e^- + h^+) \tag{2.7}$$

$$MIL - 125(Ti) + e^- \rightarrow Ti^{3+} - MIL - 125(Ti) + {}^\cdot O_2^- \tag{2.8}$$

$$Ti^{3+} - MIL - 125(Ti) + e^- + O_2 \rightarrow Ti^{4+} - MIL - 125(Ti) + {}^\cdot O_2^- \tag{2.9}$$

$${}^\cdot O_2^- + RhB - RhB^{\cdot+} \rightarrow Other\,products \rightarrow CO_2 + H_2O \tag{2.10}$$

$$h^+ + RhB - RhB^{+\cdot} \rightarrow Other\,products \rightarrow CO_2 + H_2O \tag{2.11}$$

Another hybrid material, including a Zr-based MOF and a graphitic carbon nitride (UiO-66/g-C_3N_4), was synthesized by Zhang *et al.* [129] to remove MB in a photocatalytic process. The result showed that MB was wholly degraded in 4 h under visible-light irradiation, thereby proving the efficiency of the synthesized photocatalyst. The produced photoelectrons transfer efficiently from the CB band of g-C_3N_4 to the CB band of UiO-66 due to the heterojunction between UiO-66 and g-C_3N_4. As a result, recombination of electron-hole decreases, and degradation efficiency increases.

Adding substances that act as electron acceptors such as hydrogen peroxide (H_2O_2), persulphate (PS), peroxydisulphate (PDS), and peroxymonosulphate (PMS) in the photocatalytic process can further improve photocatalytic degradation efficiency. The electron acceptors prevent recombination of photogenerated carriers by generating active hydroxyl/sulphate radicals [136–139]. Quang *et al.* [140] synthesized MIL-53(Fe) to investigate the photocatalytic degradation of MB in the presence of H_2O_2. The results showed that MB was entirely removed by MIL-53(Fe) within 30 min. Ai *et al.* [130] synthesized MIL-53(Fe) for degradation of RhB dye in the presence of H_2O_2 with a degradation efficiency of 100% under visible-light irradiation. They examined the mechanism of degradation and concluded that photogenerated electrons and Fe(III) on the surface of the MOF react with H_2O_2. Then, large amounts of hydroxyl radicals are formed that enhance the degradation of RhB dye. In another work, CuO nanoparticles were deposited on ZIF-8 surfaces by Chakraborty *et al.* [131] to investigate the photodegradation of rhodamine 6G (Rh6G) in the presence of H_2O_2 under sunlight irradiation. By conducting some different experiments, they observed that the nanocomposite containing 5 wt% of CuO content shows the best degradation efficiency by removing 99.5% of dye. In another work, a nickel-based MOF was employed as a photocatalyst for degradation of RR-120 dye by Ramachandran *et al.* [133]. This framework was chosen because of the well-oriented larger dimension of thin sheets, excellent reusability, and large surface area. It showed 89% degradation efficiency for RR-120 after 10 min in the presence of PMS. Gao *et al.* [132] reported on MIL-53(Fe) for the degradation of acid orange 7 (AO7) from aqueous solution under visible LED light and in the presence of PS. The degradation efficiency of the MIL-53(Fe) for removing AO7 in the presence of PS was 100%, compared to the value of 24% in a similar condition without using PS.

2.3.2 Photocatalytic removal of pharmaceuticals and personal care products (PPCPs) using MOFs

Pharmaceuticals and personal care products (PPCPs) are emerging contaminants that include medications, food supplements, dental care products, and cosmetics ingredients [141]. These chemicals are an important and dangerous type of contaminant and options for their removal from the aquatic environment have attracted a great deal of attention from researchers [142]. These pollutants are found at low concentrations in the environment, but extensive research has shown that in the long term, they can have very destructive effects on human health and aquatic life. PPCPs have a wide range of applications in agriculture, livestock, and human life [143]. Recently, there has been increased attention for removing this type of contaminant due to the widespread release of such contaminants into the environment and the ecological risk (i.e., creating antibiotic resistance) associated with them [144]. These contaminants, which are biologically active materials, are soluble in water [145]. PPCPs can be divided into three different groups based on their dissipation time (DT50), that is, (1) low persistence (DT = 3–7 days) such as ibuprofen and paracetamol, (2) moderate persistence (DT50 = 15–54 days) such as oxazepam, and (3) high persistence (DT50 = 119–328 days) such as clofibric acid and carbamazepine [146]. Antibiotics are an essential group of pharmaceuticals and one of the most dangerous pollutants in the environment. These drugs are widely used for both human and veterinary applications [147]. Antibiotics are discharged into the environment from various sources such as pharmaceutical industries, agricultural runoffs, hospitals, and so on. [148, 149]. However, developing efficient methods to remove antibiotics from the aqueous phase is a severe environmental challenge [150]. Among different treatment methods for removing PPCPs, photocatalytic degradation is known as an efficient and promising method due to its ability to mineralize pollutants efficiently into H_2O and CO_2 [151]. MOFs are a promising candidate for photocatalytic removal of PPCPs due to the same reasons explained in Section 2.3.1 [152]. Some recent results on PPCP degradation using MOF-based photocatalysts are summarized in Table 2.2.

Liang *et al.* [154] investigated the application of Pd@MIL-100(Fe) nanocomposite for the degradation of three typical PPCPs, ibuprofen (IBP), bisphenol A and theophylline, for the first time, in 2015. The composite was synthesized via the alcohol reduction method. The photocatalytic removal of typical PPCPs was investigated using the synthesized composite with removal efficiencies of 99.5%, 99.5%, and 64%, respectively, for theophylline, ibuprofen, and bisphenol. The introduction of Pd into the nanocomposites minimized the recombination of photogenerated electron-hole pairs. For this reason, the photoactivity of Pd@MIL-100(Fe) was reported to be much higher than that of bare MIL-100(Fe). The results also showed that 1%Pd@MIL-100(Fe) has the highest photodegradation efficiency among all the synthesized materials. Wang *et al.* [155] showed that Fe-based MOFs could be optimally used to degrade tetracycline (TC), one of the most widely used antibiotics. Fe-MIL-101 showed an excellent performance, with a TC removal efficiency of 96.6%. Due to its

Table 2.2 Recent studies on photocatalytic degradation of PPCPs.

MOF	Target Pollutant	Molecular Formula	Degradation Efficiency	Light Source	Time	References
In$_2$S$_3$/UiO−66	Tetracycline	C$_{22}$H$_{24}$N$_2$O$_8$·xH$_2$O	79.3	Vis	60	[153]
Pd@MIL−100(Fe)	Ibuprofen	C$_{13}$H$_{18}$O$_2$	99.5	Vis(H$_2$O$_2$)	150	[154]
Pd@MIL−100(Fe)	Bisphenol A	C$_{15}$H$_{16}$O$_2$	64	Vis(H$_2$O$_2$)	240	[154]
Pd@MIL−100(Fe)	Theophylline	C$_7$H$_8$N$_4$O$_2$	99.5	Vis(H$_2$O$_2$)	150	[154]
MIL−101(Fe)	Tetracycline	C$_{22}$H$_{24}$N$_2$O$_8$·xH$_2$O	96.6	Vis	180	[155]
MIL−100(Fe)/TiO$_2$	Tetracycline	C$_{22}$H$_{24}$N$_2$O$_8$·xH$_2$O	90.79	UV-Vis	10	[156]
MIL−53(Fe)	Carbamazepine	C$_{15}$H$_{12}$N$_2$O	90	Vis	60	[157]
MIL−53(Fe)	Clofibric acid	C$_{10}$H$_{11}$ClO$_3$	90	Vis	60	[157]
MIL−68(In)−NH$_2$/GO	Amoxicillin	C$_{16}$H$_{19}$N$_3$O$_5$S	93	Vis	120	[158]
MIL−68(In)−NH$_2$/GO	Amoxicillin	C$_{16}$H$_{19}$N$_3$O$_5$S	93	Vis	120	[158]
MIL−101(Fe)/TiO$_2$	Tetracycline	C$_{22}$H$_{24}$N$_2$O$_8$·xH$_2$O	92.76	Sunlight	10	[156]
AgI/UiO−66	Sulfamethoxazole	C$_{10}$H$_{11}$N$_3$O$_3$S	99.6	Vis	20	[159]
TiO$_2$@MIL−101(Cr)	Bisphenol A	C$_{15}$H$_{16}$O$_2$	99.4	Vis	240	[160]
AgFeO$_2$-graphene/Cu$_2$(BTC)$_3$	Amoxicillin	C$_{16}$H$_{19}$N$_3$O$_5$S	97	Sunlight	150	[161]
AgFeO$_2$-graphene/Cu$_2$(BTC)$_3$	Diclofenac	C$_{14}$H$_{11}$Cl$_2$NO$_2$	97	Sunlight	150	[161]
Pd-PTA-MIL-100(Fe)	Ibuprofen	C$_{13}$H$_{18}$O$_2$	99	Vis	180	[162]
Pd-PTA-MIL-100(Fe)	Theophylline	C$_7$H$_8$N$_4$O$_2$	99.5	Vis	150	[162]
CuWO$_4$/Bi$_2$S$_3$/ZIF-67	Cephalexin	C$_{16}$H$_{17}$N$_3$O$_4$S	90.1	Vis	80	[163]
CuWO$_4$/Bi$_2$S$_3$/ZIF-67	Metronidazole	C$_6$H$_9$N$_3$O$_3$	95.6	Vis	80	[163]
BiO/MIL-125(Ti)	Tetracycline	C$_{22}$H$_{24}$N$_2$O$_8$·xH$_2$O	80	Vis	120	[164]
CdS/MIL-53(Fe)	Ketorolac tromethamine	C$_{19}$H$_{24}$N$_2$O$_6$	80	Vis	330	[165]
Pt/MIL-125(Ti)/Ag	Ketoprofen	C$_{16}$H$_{14}$O$_3$	95.5	Vis	120	[166]
Ag/AgCl@MIL−88A	Ibuprofen	C$_{13}$H$_{18}$O$_2$	100	Vis	210	[167]

Metal organic frameworks for photocatalytic water treatment 51

simple synthesis, excellent photocatalytic performance, and high removal efficiency, Fe-MIL-101 can be used to degrade tetracycline and other antibiotics. In another study, MIL-100(Fe)/TiO_2 was synthesized by He et $al.$ [168] for photocatalytic degradation of TC. This composite material was chosen due to the enhanced light absorption capacity, accessibility of active sites, and electron-hole separation, which resulted in a TC removal efficiency of 90.79%. In another work, MIL-53(Fe) was successfully synthesized via a solvothermal method by Gao et $al.$ [157] and used for removing carbamazepine (CBZ) and clofibric acid (CA) from wastewater under visible-light irradiation. MIL-53(Fe) showed high photocatalytic activity and enhanced stability for the photodegradation of CBZ and CA. For both CBZ and CA compounds, photodegradation efficiency reached up to 90%. Furthermore, MIL-53(Fe) showed excellent performance for removing CBZ and CA from real municipal wastewater.

Amoxicillin (AMC) is another of the most widely used antibiotics worldwide. Removal of AMC is not easily possible with conventional wastewater treatment methods. The advanced oxidation process is used to degraded AMC [169]. For the degradation of AMC, a visible-light-driven MIL-68(In)-NH_2/GO composite was synthesized by Yang et $al.$ [158]. This composite material shows good reusability and stability. Yang et $al.$ found that the acidity of the solution has a high impact on the photodegradation efficiency. The degradation reaction at pH $=5$ showed an efficiency of 93%. Graphene oxide was added as a light sensitizer for enhancing visible-light absorption and as an electron transporter for suppressing the recombination of photogenerated carriers. He et $al.$ [156] synthesized a magnetic MIL-101(Fe)/TiO_2 for TC removal under solar-light irradiation. The composite showed a high TC degradation efficiency of 92.76% at pH $=7$ for 10 min. A visible-light responsive catalyst, AgI/UiO-66, was synthesized by Wang et $al.$ [159] for photocatalytic degradation of sulfamethoxazole (SMZ). The results showed an enhanced SMZ degradation efficiency of 99.6%, compared to the pure AgI. The SMZ degradation pathway consists of three steps: phenyl nitrification, isoxazole ring hydroxylation, and S-N bond cleavage. SMZ can eventually be converted to CO_2 and H_2O. Tang et $al.$ synthesized TiO_2@MIL-101(Cr) by a solvothermal method to investigate photocatalytic degradation of bisphenol A [160] with degradation of 99.4% in 4 h. El-Fawal et $al.$ [161] synthesized $AgFeO_2$-graphene/$Cu_2(BTC)_3$ to investigate photocatalytic degradation of amoxicillin and diclofenac (DCF) under sunlight irradiation. The degradation efficiency was measured to be 97% in 150 min. Liang et $al.$ [162] synthesized a MIL-100(Fe) impregnated with Pd-NPs-decorated phosphotungstic acid (PTA). Synergetic effects between the visible-light absorption capacity of MIL-100(Fe), electronic conductivity of the Pd NPs, and the fast electron transport of PTA resulted in excellent photocatalytic performance in the presence of H_2O_2. This hybrid photocatalyst showed a degradation efficiency of 99.5% for theophylline. Askari et $al.$ [163] synthesized $CuWO_4$/Bi_2S_3/ZIF-67 composite MOF-based material with a high surface area, excellent photocatalytic properties, enhanced chemical stability, and good reusability. The composite was used to investigate the photodegradation removal of cephalexin (CFX) and metronidazole (MTZ). The maximum degradation efficiencies were 90.1% and 95.6% for CFX and MTZ, respectively, at a constant pH of 7. Bismuth oxyiodide/MIL-125(Ti) was

52   Photocatalytic Water and Wastewater Treatment

synthesized via a one-step hydrothermal method by Jiang *et al.* [164] with a high specific surface area, good stability, reusability, and promoting visible-light absorption properties. This photocatalyst was used for the degradation of TC under visible-light and showed a degradation efficiency of 80%, which was higher than that of bare MIL-125(Ti). Chaturvedi *et al.* [165] synthesized CdS/MIL-53(Fe) via a hydrothermal method to investigate the degradation of ketorolac tromethamine (KTC). A degradation efficiency of 80% was achieved within 330 min under visible-light irradiation. Miao *et al.* [166] successfully synthesized Pt/MIL-125(Ti)/Ag by a solvothermal method to study the degradation process of ketoprofen (KP). Pt/MIL-125(Ti)/Ag photocatalyst showed 95.5% degradation of KP under visible light after 2 h. The results mentioned above, show the effectiveness of MOF-based photocatalysts for the removal of PPCPs.

2.3.3 Photocatalytic removal of heavy metals using MOFs

Nowadays, water pollution with heavy metals is one of the major concerns of the global community. Metals with a density of more than 5 g/cm^3 are heavy metals. Copper (Cu), mercury (Hg), thallium (Tl), chromium (Cr), lead (Pb), cadmium (Cd), and arsenic (As) are the heavy metals most frequently found in the environment [170, 171]. Heavy metals can cause skin damage, disruption of blood circulation, kidney damage, nervous system damage, gastrointestinal damage, asthma, chronic carcinogenic respiratory infections, and allergies [172, 173]. An increase in industrial activities and the expansion of various industries worldwide have exacerbated heavy metal toxicity. Industrial/transportation units such as mining units, coal-based power plants, solid waste disposal units, and vehicles are the most important sources of heavy metal contamination [174, 175]. Surface water can carry heavy metals over long distances and move contaminants from one point to another one. The transfer of heavy metals by surface water can lead to their bioaccumulation, causing toxicity and damage for all living entities. Heavy metals are non-biodegradable and become toxic at more than a specific concentration. Some heavy metals, such as Cd, can remain in the human body for years [176]. Since heavy metals have destructive effects on human health, it is critical to remove these pollutants from industrial wastewaters, drinking water sources, and other water sources. Different treatment methods such as electrocoagulation [177], membrane filtration [178], modified adsorption [179], microbiological remediation [180], and photoreduction are used to remove heavy metals from the environment. Compared to the other conventional purification methods, photocatalytic reduction using photocatalysts is a more effective method which does not result in any hazardous by-products [181, 182]. Several studies have been focused on developing highly efficient photocatalysts for the removal of heavy metals. Based on MOF structural features, the photosynthesis process could include three main stages: light absorption, generation of redox equivalents (i.e., electrons and holes), and reduction/oxidation half-reactions with the redox equivalents [183]. MOFs with large surface area, high density of metal sites, high chemical tunability, and organic linker diversity can be suitable candidates for photocatalytic removal of heavy metals [184]. Some recent results on CR(VI) degradation using MOF-based photocatalysts are summarized in Table 2.3.

Metal organic frameworks for photocatalytic water treatment

Cr(VI) is one of the most toxic and dangerous heavy metal cations, and many studies have been conducted on the removal of Cr(VI) or reduction of highly toxic Cr(VI) to Cr(III), which has a much lower toxicity [197, 198]. Different methods such as membrane separation, adsorption, ion exchange, and reduction have been reported for Cr(VI) removal [199]. Photocatalytic reduction of Cr(VI) from wastewater using renewable solar energy is an effective and promising method to control chromium pollution. Jing *et al.* [185] investigated MIL-68(Fe) as an active catalyst for the reduction of Cr(VI). They observed that Cr(VI) is ultimately reduced to Cr(III) within just 5 min under visible light at a constant pH of 3. Titanium-oxo clusters can facilitate the efficient transfer of the charges from excited organic linker states to the cluster. Therefore, the MOFs constructed from Ti centers, which are photo-active nodes, are good candidates for Cr(VI) reduction [200]. Wang *et al.* [193] synthesized NH_2-MIL-125(Ti) via a facile solvothermal method to study the reduction of Cr(VI) to Cr(III) under visible-light irradiation. This framework with a large surface area, decent chemical/thermal stability, and mesoporous structure showed a reduction efficiency of 97% in just 1 h. The reduction mechanism of Cr(VI) to Cr(III) in the presence of ethanol, citric acid, and ethylene diamine tetraacetic acid (EDTA) is presented in the following:

$$NH_2 - MIL - 125(Ti) + hv \rightarrow NH_2 - MIL - 125(Ti) \tag{2.12}$$

$$CH_3CH_2OH + h^+ \rightarrow CH_3CH_2OH + H_2O \rightarrow CO_2 + H_2O + Other\,products \tag{2.13}$$

$$14H^+ + Cr_2O_7^{2-} + 6e^- \rightarrow 2Cr^{3+} + 7H_2O \tag{2.14}$$

Shi *et al.* [187] synthesized the NH_2-MIL-88B(Fe) framework for the reduction of Cr(VI) under visible light. Wang *et al.* [188] synthesized ZnO@

Table 2.3 Some recent studies on photocatalytic reduction of Cr(VI) to Cr(III).

MOF	Degradation Efficiency (%)	Light Source	Time (min)	References
MIL−68(Fe)	100	UV	5	[185]
NH_2-MIL−125(Ti)	97	Vis	60	[186]
NH_2-MIL−88B(Fe)	100	Vis	45	[187]
ZnO@ZIF−8	100	Vis	80	[188]
Pd@UiO−66-NH_2	100	Vis	100	[189]
g-C_3N_4/MIL−53(Fe)	100	Vis	180	[190]
rGO/MIL−53(Fe)	100	Vis	80	[191]
MIL−68(In)	97	Vis	60	[192]
NH_2-MIL−125	80	Vis	60	[193]
Ag/AgCl@MIL−53(Fe)	100	Vis	180	[194]
Pd@MIL−101(Cr)	100	UV-Vis	60	[195]
BUC−66	98	UV	30	[196]
BUC−67	99	UV	30	[196]

54 Photocatalytic Water and Wastewater Treatment

ZIF-8 to study the photoreduction of Cr(VI) to Cr(III) with an efficiency of 100% in 80 min. Pd@UiO-66-NH$_2$ was synthesized by Shen et $al.$ as a catalyst to reduce to Cr(IV) [189]. This framework with decent photocatalytic properties could ultimately reduce all the Cr(VI) to Cr(III) under visible-light. Huang et $al.$ [190] synthesized g-C$_3$N$_4$/MIL-53(Fe) with higher photocatalytic activity compared to pure g-C$_3$N$_4$ and MIL-53(Fe). This hybrid nanocomposite showed good stability and reusability features. The reduction mechanism of Cr(VI) to Cr(III) over g-C$_3$N$_4$/MIL-53(Fe) is as follows:

$$g - C_3N_4 + hv \rightarrow h^+(g - C_3N_4) + e^- (g - C_3N_4) \tag{2.15}$$

$$MIL - 53(Fe) + hv \rightarrow h^+ \left(MIL - 53(Fe)\right) + e^- \left(MIL - 53(Fe)\right) \tag{2.16}$$

$$Cr_2O_7^{2-} + 14H^+ + 6e^- \rightarrow 2Cr^{3+} + 7H_2O \tag{2.17}$$

Liang et $al.$ [191] synthesized reduced graphene oxide (rGO)/MIL-53(Fe) for photoreduction of Cr(VI) to Cr(III) under visible-light for 80 min with a reduction efficiency of 100%. The sufficient interfacial surface contact between rGO and MIL-53(Fe) further accelerates the transfer of photogenerated electron through the structure. Furthermore, rGO effectively minimized the recombination of photogenerated electron-hole pairs, leading to an enhancement in photocatalytic performance of MIL-53(Fe). Besides, Liang et $al.$ [192] synthesized MIL-68(In) via a solvothermal method for the reduction of Cr(VI) with an efficiency of 97%. The redox process is as follows.

$$MIL - 68(In) - NH_2 \rightarrow MIL - 68(In) - NH_2 \left(h^+ + e^-\right) \tag{2.18}$$

$$14H^+ + Cr_2O_7 + 6e^- \rightarrow 2Cr^{3+} + 7H_2O \tag{2.19}$$

A highly efficient MOF-based photocatalyst, Ag/AgCl@MIL-53(Fe), was synthesized by Liu et $al.$ [194] for complete reduction of Cr(VI) under visible-light in 4 h. A more porous MOF structure, that is, MIL-101, as a host structure was impregnated with Pd particles (Pd@MIL-101(Cr)) for the photocatalytic reduction of Cr(VI) [195]. Due to the large surface area of MIL-101, its high stability under acidic conditions, and decent catalytic properties, Cr(VI) could be completely removed from the aqueous solution in the presence of formic acid. A two-dimensional Cd/Co-based metal-organic framework, BUC-66, was synthesized under hydrothermal conditions by Hong Yi et $al.$ [196] for Cr(VI) removal under UV light illumination after 30 min. The Cr(VI) reduction efficiencies reached 98% and 99% for BUC-66 and BUC-67, respectively. BUC-66 and BUC-67 were used in the photocatalytic degradation of organic dyes, that is, MO, and showed degradation efficiencies of 85% and 100%, respectively.

2.3.4 Photocatalytic degradation of pesticides and herbicides using MOFs

With the increase in world population, the request for food has been drastically increased. Because of the development of agriculture worldwide, the use of pesticides and herbicides has also been increased. Widespread use of pesticides

Metal organic frameworks for photocatalytic water treatment 55

and herbicides has caused many concerns due to their toxicity and adverse effects on human health and other living organisms. The most important and widely used herbicides are 2,4-dichlorophenoxyacetic acid (2,4-D), amitrole ($C_2H_4N_4$), ametryn ($C_9H_{17}N_5S$), bentazon ($C_{10}H_{12}N_2O_3S$), metribuzin ($C_8H_{14}N_4OS$), glyphosate ($C_3H_8NO_5P$), and atrazine ($C_8H_{14}ClN_5$). Similarly, the most important and widely used pesticides are carbaryl ($C_{12}H_{11}NO_2$), diazinon ($C_{12}H_{21}N_2O_3PS$), aldrin ($C_{12}H_8Cl_6$), secbumeton ($C_{10}H_{19}N_5O$), and oxamyl ($C_7H_{13}N_3O_3S$) [201]. Pesticides and herbicides pollute soil and water resources, causing severe problems for the health of humans and other living species [202]. There has been little research on the photocatalytic degradation of pesticides and herbicides using MOFs. Oladipo *et al.* [203] synthesized a composite material, that is, WO_3/MIL-53(Fe) framework, with a 100% removal efficiency of 2,4-D under sunlight irradiation within 4 h. They also reported on a nanohybrid composite, $AgIO_3$/MIL-53(Fe), for the degradation of methyl malathion (MLT) and chlorpyrifos (CP) as organophosphorus pesticides. Using the $AgIO_3$/MIL-53(Fe) composite material, about 90% of both MP and CP were degraded within two hours under sunlight irradiation [204]. Xue *et al.* [205] synthesized a visible-light-driven responsive photocatalyst, BiOBr/UiO-66, and applied it for the degradation of atrazine. Using this photocatalyst, the degradation efficiency of atrazine was measured to be 88%, under visible-light irradiation within 4 h, which was higher than that of pure BiOBr. Some recent results on pesticide and herbicide degradation using MOF-based photocatalysts are summarized in Table 2.4.

2.3.5 Photocatalytic degradation of other organic pollutants using MOFs

In addition to the photocatalytic removal of organic pollutants discussed above, MOFs can also be used to removed phenolic and nitroaromatics compounds. These aromatic compounds are very toxic, even at low concentrations. They are usually discharged into the environment from industrial effluents such as those from pharmaceuticals, manufacture of plastic, textile/paper, and petroleum refining. Phenol exposure damages the heart, kidney, liver, and nervous system [206]. Also, nitroaromatic compounds (NACs) are widely used

Table 2.4 Some recent studies on photocatalytic degradation of pesticides and herbicides.

MOF	Target Pollutant	Molecular Formula	Degradation Efficiency (%)	Light Source	Time (min)	References
WO_3/ MIL–53(Fe)	2,4-D	$C_8H_6Cl_2O_3$	100	Sunlight	240	[203]
$AgIO_3$/ MIL–53(Fe)	Methyl malathion	$C_7H_{13}O_6PS_2$	90	Sunlight	120	[204]
$AgIO_3$/ MIL–53(Fe)	chlorpyrifos	$C_9H_{11}Cl_3NO_3PS$	90	Sunlight	120	[204]
BiOBr/ UiO–66	Atrazine	$C_8H_{14}ClN_5$	88	Vis	240	[205]

56   **Photocatalytic Water and Wastewater Treatment**

Table 2.5 Some recent studies on photocatalytic degradation of phenolic and nitroaromatic compounds.

MOF	Target Pollutant	Molecular Formula	Degradation Efficiency (%)	Light Source	Time (min)	References
$[Zn_2(Fe-L)_2(\mu-O)-(H_2O)_2].4DMF.4H_2O$	2-chlorophenol	C_6H_5ClO	73	Vis	80	[208]
$[Zn(oba)(4-bpdh)_{0.5}]_n.1.5DMF$	phenol	C_6H_5OH	80.4	UV	120	[209]
$[Zn(oba)(4-bpdh)_{0.5}]_n.1.5DMF$	phenol	C_6H_5OH	69.3	Vis		[209]
$[Cd_{0.3}Zn_{0.7}(oba)(4-bpdh)_{0.5}]_n.1.5DMF$	phenol	C_6H_5OH	87.6	UV	120	[209]
$[Cd_{0.3}Zn_{0.7}(oba)(4-bpdh)_{0.5}]_n.1.5DMF$	phenol	C_6H_5OH	78	Vis	120	[209]
$[Ag_4(NO_3)_4(dpppda)]_n$	2,4-dinitrophenol	$C_6H_4N_2O_5$	100	UV	300	[210]
$[Ag_4(NO_3)_4(dpppda)]_n$	p-Nitrophenol	$C_6H_5NO_3$	100	UV	300	[210]
$[Ag_4(NO_3)_4(dpppda)]_n$	nitrobenzene	$C_6H_5NO_2$	100	UV	300	[210]
$Cu_2(BTC)_3@SiO_2$	phenol	C_6H_5OH	93.1	Vis	45	[211]

in various applications such as the synthesis of dyes, polymers, pesticides, and other intermediates. Most NACs can cause cancer and severe health problems for humans [207]. Therefore, the treatment of aqueous effluents containing the pollutants mentioned above is crucial. A heterometallic-organic framework, $[Zn_2(Fe-L)_2(\mu-O)-(H_2O)_2].4DMF.4H2O$, was synthesized by Li *et al.* [208], and investigated for the photocatalytic degradation of 2-chlorophenol. The results showed that 73% of 2-chlorophenol was degraded under visible light within 80 min. Masoomi *et al.* [209] synthesized two mixed metal-organic frameworks, [Zn(oba)(4-bpdh)0.5]n.1.5DMF and [Cd0.3Zn0.7(oba)(4-bpdh)0.5] n.1.5DMF. These two MOFs were studied for phenol photodegradation from an aqueous solution. The degradation efficiencies of phenol were measured to be 80.4% and 87.6% using [Zn(oba)(4-bpdh)0.5]n.1.5DMF and [Cd0.3Zn0.7(oba) (4-bpdh)0.5]n.1.5DMF, respectively. Wu *et al.* [210] used [Ag4(NO3)4(dpppda)] n (dpppda, 1,4-N,N,N',N'-tetra(diphenyl phosphanyl methyl) benzene diamine) for photocatalytic degradation of 2,4-dinitrophenol (2,4-DNP), p-Nitrophenol (PNP), and nitrobenzene. The results showed that all of the three compounds were completely degraded within 5 h under UV light irradiation. The results also indicated that the degradation reaction mainly occurs on the surface of the MOF structure. Some recent results on phenolic and nitroaromatic degradation using MOF-based photocatalysts are summarized in Table 2.5.

2.4 PARAMETERS AFFECTING THE PHOTOCATALYTIC PERFORMANCE

In addition to the type of catalyst used for a photocatalytic reaction, several operational parameters affect the performance of a photocatalytic reaction. In this regard, the effects of the most critical parameters addressed in the literature are given below.

2.4.1 Light intensity and wavelength

One of the most important factors influencing the photocatalytic reaction is light intensity. In a photocatalytic reaction, an appropriate reaction rate and electron-hole formation is required, which is dependent on the light intensity [212]. The energy of radiating light and its wavelength affect the optical and electronic properties of the catalyst. The light wavelength can be generally classified into three groups: visible, solar, or UV irradiation. With this in mind, one of the main challenges is to synthesize semiconductors with lower bandgaps that can be activated via irradiation by light having a higher wavelength (i.e., lower energy), such as visible or solar light [213]. Some MOFs, such as the MIL type, have a low bandgap and can be activated with visible-light irradiation. ZnO has a bandgap of 3.37 eV [214], and is activated with a light wavelength of less than 387 nm [215]. In contrast, MIL-101 can be activated with a light wavelength of more than 420 nm [216]. Similarly, NH2-MIL-125(Ti) can be activated with a light wavelength of more than 400 nm [217].

2.4.2 Photocatalyst loading

The photocatalytic degradation rate usually increases with increasing photocatalyst amount due to the higher generation of reactive radicals. However, in very concentrated solutions, the degradation rate decreases due to screening effects and light scattering [218]. Therefore, finding the optimum catalyst loading is essential for enhanced photocatalytic degradation efficiency [219].

2.4.3 Dissolved oxygen

Dissolved oxygen can directly affect the formation of intermediate species during photocatalytic reactions, stabilizing intermediate radicals and avoiding charge recombination effects. Furthermore, oxygen molecules play a direct role in the production of reactive oxygen species (ROS) responsible for the degradation of pollutants.

2.4.4 The effect of pH

pH is one of the critical and influential parameters in the conduction of photocatalytic reactions. pH can directly affect the reaction efficiency by modifying the photocatalyst surface charge and affecting particle self-aggregations [220]. Electrostatic interactions between the photocatalyst surface, charged radicals, solvent molecules, and the substrate are dependent on the solution's acidity [221]. pH affects the point of zero charge (pH_{PZC}) of the photocatalyst. At pH values near the pH_{PZC}, due to the absence of electrostatic forces, the interactions between the photocatalyst surface and solved species are negligible. However, at pH less than the pH_{PZC}, the photocatalyst surface is positively charged, and an electrostatic attraction occurs with the negatively charged compounds, which can increase the absorption capacity of photocatalysts and enhance the degradation efficiency. On the other hand, at pH values more than the pH_{PZC}, the photocatalyst surface will be charged negatively, causing repulsion between the MOF surface and the anionic compounds [32].

2.4.5 Initial concentration of contaminants

Increasing the initial concentration of contaminants leads to a decrease in the photodegradation reaction efficiency, which could be due to a decrease in the photonic efficiency and saturation of the photocatalytic active sites by the adsorbed molecules [222]. Besides, the concentration of the intermediate species increases with increasing initial concentration of the contaminant. The saturation of the photocatalytic active sites by the intermediate species can lead to a decrease in the photocatalytic reaction rate and the degradation efficiency [223].

2.5 CONCLUSIONS AND OUTLOOK

Photocatalytic degradation of pollutants is an emerging and promising method of wastewater treatment. Recently, MOFs, as a group of highly porous materials with crystalline nature, have been used in various applications. MOF structures, with unique properties, including unprecedently high surface area and chemical tunability are suitable catalysts for the removal of water contaminants.

Combining MOFs with semiconductors and nanomaterials can further improve their performance in the photodegradation of pollutants. Degradation of water pollutants under UV light illumination has been discussed in many studies in the literature. More effort is required in the synthesis of novel structures which are capable of degrading contaminants under visible or sunlight irradiation. This is where MOFs can play a significant role.

Dye removal has been studied only for a few of the more well-known dyes to date, and the degradation of other dyes should also be considered. Besides, more research is required to evaluate the degradation of other pollutants, including pesticides, herbicides, and personal care products and the reduction of heavy metals. MOFs have recently started to be used for the degradation of some of the latter contaminants. However, it is essential to further study the targeted synthesis methods and novel MOF structures in photocatalytic applications, in order to achieve enhanced photocatalytic performance.

Most studies in this area, to date, have been on the lab-scale application of MOFs for photocatalytic degradation of water contaminants. However, it is also vital to scale-up water treatment processes using MOFs for industrial applications. Such processes require more facile, reliable and economically viable material synthesis methods. In addition, for the catalysts used in such large-scale processes, stability is another essential factor to be considered. Therefore, developing highly stable MOF structures with adequate light-harvesting properties to be used in aqueous and acidic environments, is in the forefront of research in this field. Recycling of MOF-based catalysts is another important parameter for industrial applications which needs to be considered. Overall, in this chapter, we showed that a number of different MOFs have recently been applied for the photocatalytic degradation of various organic pollutants in aquatic systems. We also reviewed the degradation mechanism of some pollutants using MOF-type photocatalysts. This chapter reveals that MOFs are promising candidates as catalysts for the photocatalytic removal of different contaminants from water.

REFERENCES

1. Gopinath KP, Madhav NV, Krishnan A, Malolan R, Rangarajan G. Present applications of titanium dioxide for the photocatalytic removal of pollutants from water: A review. *Journal of Environmental Management* 2020; **270**: 110906.
2. Mon M, Bruno R, Ferrando-Soria J, Armentano D, Pardo E. Metal–organic framework technologies for water remediation: towards a sustainable ecosystem. *Journal of Materials Chemistry A* 2018; **6**: 4912–47.
3. Milly PCD, Betancourt J, Falkenmark M, *et al.* Stationarity Is Dead: Whither Water Management? *Science* 2008; **319**: 573.
4. Vörösmarty CJ, McIntyre PB, Gessner MO, Dudgeon D, Prusevich A, Green P, Glidden S, Bunn SE, Sullivan CA, Liermann CR, Davies PM. Global threats to human water security and river biodiversity. *Nature* 2010; **467**: 555–61.
5. Schwarzenbach RP, Escher BI, Fenner K, Hofstette TB, Johnson CA, von Gunten Urs, Wehrli B. The Challenge of Micropollutants in Aquatic Systems. *Science* 2006; **313**: 1072.
6. Gupta VK, Ali I, Saleh TA, Nayak A, Agarwal S. Chemical treatment technologies for waste-water recycling—an overview. *RSC Advances* 2012; **2**: 6380–8.
7. Savage N, Diallo MS. Nanomaterials and Water Purification: Opportunities and Challenges. *Journal of Nanoparticle Research* 2005; **7**: 331–42.
8. Shannon MA, Bohn PW, Elimelech M, Georgiadis JG, Mariñas BJ, Mayes AM. Science and technology for water purification in the coming decades. *Nature* 2008; **452**: 301–10.
9. Verma AK, Dash RR, Bhunia P. A review on chemical coagulation/flocculation technologies for removal of colour from textile wastewaters. *Journal of Environmental Management* 2012; **93**: 154–68.
10. Gao Q, Xu J, Bu X-H. Recent advances about metal–organic frameworks in the removal of pollutants from wastewater. *Coordination Chemistry Reviews* 2019; **378**: 17–31.
11. Justicia I, Ordejón P, Canto G, Mozos JL, Fraxedas J, Battiston GA, Gerbasi R, Figueras A. Designed Self-Doped Titanium Oxide Thin Films for Efficient Visible-Light Photocatalysis. *Advanced Materials* 2002; **14**: 1399–402.
12. Lee G-J, Wu JJ. Recent developments in ZnS photocatalysts from synthesis to photocatalytic applications – A review. *Powder Technology* 2017; **318**: 8–22.
13. Wang P, Huang B, Zhang Q, Zhang X, Qin X, Dai Y, Zhan J, Yu J, Liu H, Lou Z. Highly Efficient Visible Light Plasmonic Photocatalyst Ag@Ag(Br,I). *Chemistry – A European Journal* 2010; **16**: 10042–7.
14. Jun W, Wang J, Rong W, Yan-Fei S, Yu-Feng Z, Ji-Kang J. Synthesis and characterization of zinc sulfide hollow microspheres. *Powder Diffraction - POWDER DIFFR* 2009; 24.
15. Mondal C, Singh A, Sahoo R, Sasmal AK, Negishi Y, Pal T. Preformed ZnS nanoflower prompted evolution of CuS/ZnS p–n heterojunctions for exceptional visible-light driven photocatalytic activity. *New Journal of Chemistry* 2015; **39**: 5628–35.
16. Hussain MZ, Schneemann A, Fischer RA, Zhu Y, Xia Y. MOF Derived Porous ZnO/C Nanocomposites for Efficient Dye Photodegradation. *ACS Applied Energy Materials* 2018; **1**: 4695–707.
17. Farha OK, Özgür Yazaydın A, Eryazici I, Malliakas CD, Hauser BG, Kanatzidis MG, Nguyen ST, Snurr RQ, Hupp JT. De novo synthesis of a metal–organic framework material featuring ultrahigh surface area and gas storage capacities. *Nat Chem* 2010; **2**: 944–8.
18. Gao W-Y, Chrzanowski M, Ma S. Metal–metalloporphyrin frameworks: a resurging class of functional materials. *Chemical Society Reviews* 2014; **43**: 5841–66.

19. Horcajada P, Gref R, Baati T, Allan PK, Maurin G, Couvreur P, Férey G, Morris RE, Serre C. Metal–Organic Frameworks in Biomedicine. *Chemical Reviews* 2012; **112**: 1232–68.
20. Murray LJ, Dincă M, Long JR. Hydrogen storage in metal–organic frameworks. *Chemical Society Reviews* 2009; **38**: 1294–314.
21. Van de Voorde B, Bueken B, Denayer J, De Vos D. Adsorptive separation on metal–organic frameworks in the liquid phase. *Chemical Society Reviews* 2014; **43**: 5766–88.
22. Jiang D, Xu P, Wang H, Zeng G, Huang D, Chen M, Lai C, Zhang C, Wan J, Xue W. Strategies to improve metal organic frameworks photocatalyst's performance for degradation of organic pollutants. *Coordination Chemistry Reviews* 2018; **376**: 449–66.
23. Akpan UG, Hameed BH. Parameters affecting the photocatalytic degradation of dyes using TiO2-based photocatalysts: A review. *Journal of Hazardous Materials* 2009; **170**: 520–9.
24. Chong MN, Jin B, Chow CWK, Saint C. Recent developments in photocatalytic water treatment technology: *A review. Water Res* 2010; **44**: 2997–3027.
25. Dhaka S, Kumar R, Deep A, Kurade MB, Ji S-W, Jeon B-H. Metal–organic frameworks (MOFs) for the removal of emerging contaminants from aquatic environments. *Coordination Chemistry Reviews* 2019; **380**: 330–52.
26. Gautam S, Agrawal H, Thakur M, Akbari A, Sharda H, Kaur R, Amini M. Metal oxides and metal organic frameworks for the photocatalytic degradation: A review. *Journal of Environmental Chemical Engineering* 2020; **8**: 103726.
27. Karthikeyan C, Arunachalam P, Ramachandran K, Al-Mayouf AM, Karuppuchamy S. Recent advances in semiconductor metal oxides with enhanced methods for solar photocatalytic applications. *Journal of Alloys and Compounds* 2020; **828**: 154281.
28. Kronawitter C, Kiriakidis G. An overview of photocatalytic materials. *Journal of Materiomics* 2017; 1–2.
29. Salim HAM, Idrees SA, Rashid RA, Mohammed AA, Simo SM, Khalo IS. Photocatalytic degradation of Toluidine Blue Dye in Aqueous Medium Under Fluorescent Light 2018 International Conference on Advanced Science and Engineering (ICOASE), 2018: 384–8.
30. Hopfield JJ. On the energy dependence of the absorption constant and photoconductivity near a direct band gap. *Journal of Physics and Chemistry of Solids* 1961; **22**: 63–72.
31. Serpone N, Emeline AV. Semiconductor Photocatalysis – Past, Present, and Future Outlook. *The Journal of Physical Chemistry Letters* 2012; **3**: 673–7.
32. Bedia J, Muelas-Ramos V, Peñas-Garzón M, Gómez-Avilés A, Rodríguez JJ, Belver C. A Review on the Synthesis and Characterization of Metal Organic Frameworks for Photocatalytic Water Purification. *Catalysts* 2019; **9**: 52.
33. Gusain R, Gupta K, Joshi P, Khatri OP. Adsorptive removal and photocatalytic degradation of organic pollutants using metal oxides and their composites: A comprehensive review. *Advances in Colloid and Interface Science* 2019; **272**: 102009.
34. Wang Q, Gao Q, Al-Enizi AM, Nafady A, Ma S. Recent advances in MOF-based photocatalysis: environmental remediation under visible light. *Inorganic Chemistry Frontiers* 2020; **7**: 300–39.
35. Konstantinou IK, Albanis TA. TiO2-assisted photocatalytic degradation of azo dyes in aqueous solution: kinetic and mechanistic investigations: A review. *Applied Catalysis B: Environmental* 2004; **49**: 1–14.

Metal organic frameworks for photocatalytic water treatment   61

36. Rauf MA, Meetani MA, Hisaindee S. An overview on the photocatalytic degradation of azo dyes in the presence of TiO2 doped with selective transition metals. *Desalination* 2011; **276**: 13–27.
37. Fagan R, McCormack DE, Dionysiou DD, Pillai SC. A review of solar and visible light active TiO2 photocatalysis for treating bacteria, cyanotoxins and contaminants of emerging concern. *Materials Science in Semiconductor Processing* 2016; **42**: 2–14.
38. Kumar SG, Rao KSRK. Comparison of modification strategies towards enhanced charge carrier separation and photocatalytic degradation activity of metal oxide semiconductors (TiO2, WO3 and ZnO). *Applied Surface Science* 2017; **391**: 124–48.
39. Wang C, Lin H, Xu Z, Cheng H, Zhang C. One-step hydrothermal synthesis of flowerlike MoS2/CdS heterostructures for enhanced visible-light photocatalytic activities. *RSC Advances* 2015; **5**: 15621–6.
40. Wen J, Xie J, Chen X, Li X. A review on g-C3N4-based photocatalysts. *Applied Surface Science* 2017; **391**: 72–123.
41. Zhang C, Li Y, Shuai D, Shen Y, Xiong W, Wang L. Graphitic carbon nitride (g-C3N4)-based photocatalysts for water disinfection and microbial control: A review. *Chemosphere* 2019; **214**: 462–79.
42. Huang Y, Fan W, Long B, Li H, Zhao F, Liu Z, Tong Y, Ji H. Visible light Bi2S3/Bi2O3/Bi2O2CO3 photocatalyst for effective degradation of organic pollutions. *Applied Catalysis B: Environmental* 2016; **185**: 68–76.
43. Madhusudan P, Ran J, Zhang J, Yu J, Liu G. Novel urea assisted hydrothermal synthesis of hierarchical BiVO4/Bi2O2CO3 nanocomposites with enhanced visible-light photocatalytic activity. *Applied Catalysis B: Environmental* 2011; **110**: 286–95.
44. Fujishima A, Honda K. Electrochemical photolysis of water at a semiconductor electrode. *Nature* 1972; **238**: 37–8.
45. Xu M, Gao Y, Moreno EM, Kunst M, Muhler M, Wang Y, Idriss H, Wöll C. Photocatalytic activity of bulk TiO2 anatase and rutile single crystals using infrared absorption spectroscopy. *Phys Rev Lett* 2011; **106**: 138302.
46. Banerjee S, Pillai SC, Falaras P, O'Shea KE, Byrne JA, Dionysiou DD. New Insights into the Mechanism of Visible Light Photocatalysis. *The Journal of Physical Chemistry Letters* 2014; **5**: 2543–54.
47. Nagalakshmi MCK, Anusuya N, Chidambaram B, Muthuramalingam J, Subbian K. Synthesis of TiO2 nanofiber for photocatalytic and antibacterial applications. *Journal of Materials Science: Materials in Electronics* 2017; 15915–15920.
48. Pan L, Huang H, Lim CK, Hong QY, Tse MS, Tan OK. TiO2 rutile–anatase core–shell nanorod and nanotube arrays for photocatalytic applications. *RSC Advances* 2013; **3**: 3566–71.
49. Testino A, Bellobono IR, Buscaglia V, Canevali C, D'Arienzo M, Polizzi S, Scotti R, Morazzoni F. Optimizing the Photocatalytic Properties of Hydrothermal TiO2 by the Control of Phase Composition and Particle Morphology. A Systematic Approach. *Journal of the American Chemical Society* 2007; **129**: 3564–75.
50. Colbeau-Justin C, Kunst MDH. Structural Influence on Charge-Carrier Lifetimes in TiO2 Powders Studied by Microwave Absorption. *Journal of Materials Science* 2003; 2429–2437.
51. Kafizas A, Wang X, Pendlebury SR, Barnes P, Ling M, Sotelo-Vazquez C, Quesada-Cabrera R, Li SR, Parkin IP, Durrant JR. Where Do Photogenerated Holes Go in Anatase:Rutile TiO2? A Transient Absorption Spectroscopy Study of Charge Transfer and Lifetime. *The Journal of Physical Chemistry A* 2016; **120**: 715–23.

52. Pellegrino F, Pellutiè L, Sordello F, Minero C, Ortel E, Hodoroaba V-D, Maurino V. Influence of agglomeration and aggregation on the photocatalytic activity of TiO2 nanoparticles. *Applied Catalysis B: Environmental* 2017; **216**: 80–7.
53. Wang M, Cai L, Wang Y, Zhou F, Xu K, Tao X, Chai Y Graphene-Draped Semiconductors for Enhanced Photocorrosion Resistance and Photocatalytic Properties. *Journal of the American Chemical Society* 2017; **139**: 4144–51.
54. Zhang M, Bosch M, Gentle Iii T, Zhou H-C. Rational design of metal–organic frameworks with anticipated porosities and functionalities. *CrystEngComm* 2014; **16**: 4069–83.
55. Doonan CJ, Sumby CJ. Metal–organic framework catalysis. *CrystEngComm* 2017; **19**: 4044–8.
56. Hirscher M, Panella B. Hydrogen storage in metal–organic frameworks. *Scripta Materialia* 2007; **56**: 809–12.
57. Lee J, Farha OK, Roberts J, Scheidt KA, Nguyen ST, Hupp JT. Metal–organic framework materials as catalysts. *Chemical Society Reviews* 2009; **38**: 1450–9.
58. Ma F-J, Liu S-X, Sun C-Y, Liang D-D, Ren G-J, Wei F, Chen Y-G, Su Z-M. A Sodalite-Type Porous Metal–Organic Framework with Polyoxometalate Templates: Adsorption and Decomposition of Dimethyl Methylphosphonate. *Journal of the American Chemical Society* 2011; **133**: 4178–81.
59. Abánades Lázaro I, Forgan RS. Application of zirconium MOFs in drug delivery and biomedicine. *Coordination Chemistry Reviews* 2019; **380**: 230–59.
60. Lin X, Jia J, Hubberstey P, Schröder M, Champness NR. Hydrogen storage in metal–organic frameworks. *CrystEngComm* 2007; **9**: 438–48.
61. Lu G, Hupp JT. Metal–Organic Frameworks as Sensors: A ZIF-8 Based Fabry-Pérot Device as a Selective Sensor for Chemical Vapors and Gases. *Journal of the American Chemical Society* 2010; **132**: 7832–3.
62. Butova VV, Soldatov MA, Guda AA, Lomachenko KA, Lamberti C. Metal-organic frameworks: structure, properties, methods of synthesis and characterization. *Turpion publications* 2016; **85**: 280.
63. Devic T, Serre C. High valence 3p and transition metal based MOFs. *Chemical Society Reviews* 2014; **43**: 6097–115.
64. Cavka JH, Jakobsen S, Olsbye U, Guillou N, Lamberti C, Bordiga S, Lillerud KP. A New Zirconium Inorganic Building Brick Forming Metal Organic Frameworks with Exceptional Stability. *Journal of the American Chemical Society* 2008; **130**: 13850–1.
65. Gomes Silva C, Luz I, Llabrés i Xamena FX, Corma A, García H. Water stable Zr-benzenedicarboxylate metal-organic frameworks as photocatalysts for hydrogen generation. *Chemistry* 2010; **16**: 11133–8.
66. Fu Y, Sun D, Chen Y, Huang R, Ding Z, Fu X, Li Z. An Amine-Functionalized Titanium Metal–Organic Framework Photocatalyst with Visible-Light-Induced Activity for CO2 Reduction. *Angewandte Chemie* 2012; **51**: 3364–7.
67. Horiuchi Y, Toyao T, Saito M, Mochizuki K, Iwata M, Higashimura H, Anpo M, Matsuoka M. Visible-Light-Promoted Photocatalytic Hydrogen Production by Using an Amino-Functionalized Ti(IV) Metal–Organic Framework. *The Journal of Physical Chemistry C* 2012; **116**: 20848–53.
68. Laurier KG, Vermoortele F, Ameloot R, De Vos DE, Hofkens J, Roeffaers MB. Iron(III)-based metal-organic frameworks as visible light photocatalysts. *J Am Chem Soc* 2013; **135**: 14488–91.
69. Silva CG, Corma A, García H. Metal–organic frameworks as semiconductors. *Journal of Materials Chemistry* 2010; **20**: 3141–56.

Metal organic frameworks for photocatalytic water treatment   63

70. Zhou T, Du Y, Borgna A, Hong J, Wang Y, Han J, Zhang W, Xu R. Post-synthesis modification of a metal–organic framework to construct a bifunctional photocatalyst for hydrogen production. *Energy & Environmental Science* 2013; **6**: 3229–34.
71. Dhakshinamoorthy A, Li Z, Garcia H. Catalysis and photocatalysis by metal organic frameworks. *Chemical Society Reviews* 2018; **47**: 8134–72.
72. Hasan Z, Jhung SH. Removal of hazardous organics from water using metal-organic frameworks (MOFs): Plausible mechanisms for selective adsorptions. *Journal of Hazardous Materials* 2015; **283**: 329–39.
73. Alvaro M, Carbonell E, Ferrer B, Llabrés i Xamena FX, Garcia H. Semiconductor behavior of a metal-organic framework (MOF). *Chemistry* 2007; **13**: 5106–12.
74. Qin Y, Hao M, Li Z. Chapter 17 - Metal–organic frameworks for photocatalysis. In: Yu J, Jaroniec M , Jiang C, eds. *Interface Science and Technology*, Elsevier, 2020: 541–79.
75. Li D, Xu H-Q, Jiao L, Jiang H-L. Metal-organic frameworks for catalysis: State of the art, challenges, and opportunities. *EnergyChem* 2019; **1**: 100005.
76. Zhang T, Lin W. Metal–organic frameworks for artificial photosynthesis and photocatalysis. *Chemical Society Reviews* 2014; **43**: 5982–93.
77. Kaur H, Kumar A, Koner RR, Krishnan V. Chapter 6 - Metal-organic frameworks for photocatalytic degradation of pollutants. In: Singh P, Borthakur A, Mishra PK , Tiwary D, eds. *Nano-Materials as Photocatalysts for Degradation of Environmental Pollutants*, Elsevier, 2020: 91–126.
78. Nasalevich MA, van der Veen M, Kapteijn F, Gascon J. Metal–organic frameworks as heterogeneous photocatalysts: advantages and challenges. *CrystEngComm* 2014; **16**: 4919–26.
79. Wen M, Mori K, Kuwahara Y, An T, Yamashita H. Design of Single-Site Photocatalysts by Using Metal–Organic Frameworks as a Matrix. *Asian chemical editorial society (ACES)* 2018; **13**: 1767–79.
80. Zhang X, Wang J, Dong X-X, Lv Y-K. Functionalized metal-organic frameworks for photocatalytic degradation of organic pollutants in environment. *Chemosphere* 2020; **242**: 125144.
81. Toyao T, Saito M, Dohshi S, Mochizuki K, Iwata M, Higashimura H, Horiuchi Y, Matsuoka M. Development of a Ru complex-incorporated MOF photocatalyst for hydrogen production under visible-light irradiation. *Chemical Communications* 2014; **50**: 6779–81.
82. Lin C-K, Zhao D, Gao W-Y, Yang Z, Ye J, Xu T, Ge Q, Ma S, Liu D-J. Tunability of Band Gaps in Metal–Organic Frameworks. *Inorganic Chemistry* 2012; **51**: 9039–44.
83. Lin W, Frei H. Photochemical CO2 Splitting by Metal-to-Metal Charge-Transfer Excitation in Mesoporous ZrCu(I)-MCM-41 Silicate Sieve. *Journal of the American Chemical Society* 2005; **127**: 1610–1.
84. Tsuji I, Kato H, Kobayashi H, Kudo A. Photocatalytic H2 Evolution Reaction from Aqueous Solutions over Band Structure-Controlled (AgIn)xZn2(1-x)S2 Solid Solution Photocatalysts with Visible-Light Response and Their Surface Nanostructures. *Journal of the American Chemical Society* 2004; **126**: 13406–13.
85. Sun D, Liu W, Qiu M, Zhang Y, Li Z. Introduction of a mediator for enhancing photocatalytic performance via post-synthetic metal exchange in metal–organic frameworks (MOFs). *Chemical Communications* 2015; **51**: 2056–9.
86. Nivetha R, Gothandapani K, Raghavan V, Jacob G, Sellappan R, Bhardwaj P, Pitchaimuthu S, Kannan ANM, Jeong SK, Grace AN. Highly Porous MIL-100(Fe) for the Hydrogen Evolution Reaction (HER) in Acidic and Basic Media. *ACS Omega* 2020; **5**: 18941–9.

87. Ghattas A-K, Fischer F, Wick A, Ternes TA. Anaerobic biodegradation of (emerging) organic contaminants in the aquatic environment. *Water Res* 2017; **116**: 268–95.
88. Grandclément C, Seyssiecq I, Piram A, Wong-Wah-Chung P, Vanot G, Tiliacos N, Roche N, Doumenq P. From the conventional biological wastewater treatment to hybrid processes, the evaluation of organic micropollutant removal: A review. *Water Res* 2017; **111**: 297–317.
89. Dias EM, Petit C. Towards the use of metal–organic frameworks for water reuse: a review of the recent advances in the field of organic pollutants removal and degradation and the next steps in the field. *Journal of Materials Chemistry A* 2015; **3**: 22484–506.
90. Chowdhury MF, Khandaker S, Sarker F, Islam A, Rahman MT, Awual MR. Current treatment technologies and mechanisms for removal of indigo carmine dyes from wastewater: A review. *Journal of Molecular Liquids* 2020; **318**: 114061.
91. Dawood S, Sen TK, Phan CJWA, Pollution S. Synthesis and Characterisation of Novel-Activated Carbon from Waste Biomass Pine Cone and Its Application in the Removal of Congo Red Dye from Aqueous Solution by Adsorption. *Water, Air, & Soil Pollution* 2014; **225**: 1818.
92. Islam MA, Ali I, Karim SMA, Hossain Firoz MS, Chowdhury A-N, Morton DW, Angove MJ. Removal of dye from polluted water using novel nano manganese oxide-based materials. *Journal of Water Process Engineering* 2019; **32**: 100911.
93. Hasanpour M, Hatami M. Photocatalytic performance of aerogels for organic dyes removal from wastewaters: Review study. *Journal of Molecular Liquids* 2020; **309**: 113094.
94. Pereira L, Alves M. Dyes—Environmental Impact and Remediation. *Environmental Protection Strategies for Sustainable Development* 2012: 111–62.
95. Yagub MT, Sen TK, Afroze S, Ang HM. Dye and its removal from aqueous solution by adsorption: a review. *Adv Colloid Interface Sci* 2014; **209**: 172–84.
96. Gupta VK, Kumar R, Nayak A, Saleh TA, Barakat MA. Adsorptive removal of dyes from aqueous solution onto carbon nanotubes: a review. *Adv Colloid Interface Sci* 2013; **193–194**: 24–34.
97. Reddy CV, Reddy KR, Harish VVN, Shim J, Shankar MV, Shetti NP, Aminabhavi TM. Metal-organic frameworks (MOFs)-based efficient heterogeneous photocatalysts: Synthesis, properties and its applications in photocatalytic hydrogen generation, CO2 reduction and photodegradation of organic dyes. *International Journal of Hydrogen Energy* 2020; **45**: 7656–79.
98. Zhang L-H, Zhu Y, Lei B-R, Li Y, Zhu W, Li Q. Trichromatic dyes sensitized HKUST-1 (MOF-199) as scavenger towards reactive blue 13 via visible-light photodegradation. *Inorganic Chemistry Communications* 2018; **94**: 27–33.
99. Jing H-P, Wang C-C, Zhang Y-W, Wang P, Li R. Photocatalytic degradation of methylene blue in ZIF-8. *RSC Advances* 2014; **4**: 54454–62.
100. Wang C-C, Zhang Y-Q, Zhu T, Wang P, Gao S-J. Photocatalytic degradation of methylene blue and methyl orange in a Zn(II)-based Metal–Organic Framework. *Desalination and Water Treatment* 2016; **57**: 17844–51.
101. Bala S, Bhattacharya S, Goswami A, Adhikary A, Konar S, Mondal R. Designing Functional Metal–Organic Frameworks by Imparting a Hexanuclear Copper-Based Secondary Building Unit Specific Properties: Structural Correlation With Magnetic and Photocatalytic Activity. *Crystal Growth & Design* 2014; **14**: 6391–8.
102. Mahmoodi NM, Abdi J. Nanoporous metal-organic framework (MOF-199): Synthesis, characterization and photocatalytic degradation of Basic Blue 41. *Microchemical Journal* 2019; **144**: 436–42.
103. Masoomi MY, Bagheri M, Morsali A. High efficiency of mechanosynthesized Zn-based metal–organic frameworks in photodegradation of congo red under UV and visible light. *RSC Advances* 2016; **6**: 13272–7.

104. Wang A, Zhou Y, Wang Z, Chen M, Sun L, Liu X. Titanium incorporated with UiO-66(Zr)-type Metal–Organic Framework (MOF) for photocatalytic application. *RSC Advances* 2016; **6**: 3671–9.
105. Du JJ, Yuan YP, Sun JX, Peng FM, Jiang X, Qiu LG, Xie AJ, Shen YH, Zhu JF. New photocatalysts based on MIL-53 metal-organic frameworks for the decolorization of methylene blue dye. *J Hazard Mater* 2011; **190**: 945–51.
106. Xu W-T, Ma L, Ke F, Peng F-M, Xu G-S, Shen Y-H, Zhu J-F, Qiu L-G, Yuan Y-P. Metal–organic frameworks MIL-88A hexagonal microrods as a new photocatalyst for efficient decolorization of methylene blue dye. *Dalton Transactions* 2014; **43**: 3792–8.
107. Guesh K, Caiuby CAD, Mayoral Á, Díaz-García M, Díaz I, Sanchez-Sanchez M. Sustainable Preparation of MIL-100(Fe) and Its Photocatalytic Behavior in the Degradation of Methyl Orange in Water. *Crystal Growth & Design* 2017; **17**: 1806–13.
108. Gao J, Miao J, Li P-Z, Teng WY, Yang L, Zhao Y, Liu B, Zhang Q. A p-type Ti(iv)-based metal–organic framework with visible-light photo-response. *Chemical Communications* 2014; **50**: 3786–8.
109. Kumar Paul A, Madras G, Natarajan S. Adsorption–desorption and photocatalytic properties of inorganic–organic hybrid cadmium thiosulfate compounds. *Physical Chemistry Chemical Physics* 2009; **11**: 11285–96.
110. Zhang CY, Ma WX, Wang MY, Yang XJ, Xu XY. Structure, photoluminescent properties and photocatalytic activities of a new Cd(II) metal–organic framework. *Spectrochimica Acta Part A: Molecular and Biomolecular Spectroscopy* 2014; **118**: 657–62.
111. Yang H, He X-W, Wang F, Kang Y, Zhang J. Doping copper into ZIF-67 for enhancing gas uptake capacity and visible-light-driven photocatalytic degradation of organic dye. *Journal of Materials Chemistry* 2012; **22**: 21849–51.
112. Mahata P, Madras G, Natarajan S. Novel Photocatalysts for the Decomposition of Organic Dyes Based on Metal-Organic Framework Compounds. *The Journal of Physical Chemistry B* 2006; **110**: 13759–68.
113. Dinh Du P, Thanh H, to tc, Thang H, Tinh M, Tuyen T, Thai Hoa T, Khieu D. Metal-Organic Framework MIL-101: Synthesis and Photocatalytic Degradation of Remazol Black B Dye. *Journal of Nanomaterials* 2019; **2019**: 1–15.
114. Ahmadpour N, Sayadi MH, Homaeigohar S. A hierarchical Ca/TiO2/NH2-MIL-125 nanocomposite photocatalyst for solar visible light induced photodegradation of organic dye pollutants in water. *RSC Advances* 2020; **10**: 29808–20.
115. Mahmoodi NM, Taghizadeh A, Taghizadeh M, Abdi J. In situ deposition of Ag/AgCl on the surface of magnetic metal-organic framework nanocomposite and its application for the visible-light photocatalytic degradation of Rhodamine dye. *Journal of Hazardous Materials* 2019; **378**: 120741.
116. Wu Y, Luo H, Zhang L. Pd nanoparticles supported on MIL-101/reduced graphene oxide photocatalyst: an efficient and recyclable photocatalyst for triphenylmethane dye degradation. *Environmental Science and Pollution Research* 2015; **22**: 17238–43.
117. Fazaeli R, Aliyan H, Banavandi RS. Sunlight assisted photodecolorization of malachite green catalyzed by MIL-101/graphene oxide composites. *Russian Journal of Applied Chemistry* 2015; **88**: 169–77.
118. Bai Y, Zhang S, Feng S, Zhu M, Ma S. The first ternary Nd-MOF/GO/Fe3O4 nanocomposite exhibiting an excellent photocatalytic performance for dye degradation. *Dalton Transactions* 2020; **49**: 10745–54.
119. Huang L, Liu B. Synthesis of a novel and stable reduced graphene oxide/MOF hybrid nanocomposite and photocatalytic performance for the degradation of dyes. *RSC Advances* 2016; **6**: 17873–9.

120. Sha Z, Sun J, On Chan HS, Jaenicke S, Wu J. Bismuth tungstate incorporated zirconium metal–organic framework composite with enhanced visible-light photocatalytic performance. *RSC Advances* 2014; **4**: 64977–84.
121. Chi L, Xu Q, Liang X, Wang J, Su X. Iron-Based Metal–Organic Frameworks as Catalysts for Visible Light-Driven Water Oxidation. *Small* 2016; **12**: 1351–8.
122. Ding J, Yang Z, He C, Tong X, Li Y, Niu X, Zhang H. UiO-66(Zr) coupled with Bi2MoO6 as photocatalyst for visible-light promoted dye degradation. *Journal of Colloid and Interface Science* 2017; **497**: 126–33.
123. Yang J, Niu X, An S, Chen W, Wang J, Liu W. Facile synthesis of Bi2MoO6–MIL-100(Fe) metal–organic framework composites with enhanced photocatalytic performance. *RSC Advances* 2017; **7**: 2943–52.
124. Zhao S, Chen Z, Shen J, Qu Y, Wang B, Wang X. Enhanced photocatalytic performance of BiOBr/NH2-MIL-125(Ti) composite for dye degradation under visible light. *Dalton Transactions* 2016; **45**: 17521–9.
125. Han Y, Bai C, Zhang L, Wu J, Meng H, Xu J, Xu Y, Liang Z, Zhang X. A facile strategy for fabricating AgI–MIL-53(Fe) composites: superior interfacial contact and enhanced visible light photocatalytic performance. *New Journal of Chemistry* 2018; **42**: 3799–807.
126. Shen L, Huang L, Liang S, Liang R, Qin N, Wu L. Electrostatically derived self-assembly of NH2-mediated zirconium MOFs with graphene for photocatalytic reduction of Cr(vi). *RSC Advances* 2014; **4**: 2546–9.
127. Zhou T, Zhang G, Zhang H, Yang H, Ma P, Li X, Qiu X, Liu G. Highly efficient visible-light-driven photocatalytic degradation of rhodamine B by a novel Z-scheme Ag3PO4/MIL-101/NiFe2O4 composite. *Catalysis Science & Technology* 2018; **8**: 2402–16.
128. Wang C, Cao M, Wang P, Ao Y. Preparation, characterization of CdS-deposited graphene–carbon nanotubes hybrid photocatalysts with enhanced photocatalytic activity. *Materials Letters* 2013; **108**: 336–9.
129. Zhang Y, Zhou J, Feng Q, Chen X, Hu Z. Visible light photocatalytic degradation of MB using UiO-66/g-C3N4 heterojunction nanocatalyst. *Chemosphere* 2018; **212**: 523–32.
130. Ai L, Zhang C, Li L, Jiang J. Iron terephthalate metal–organic framework: Revealing the effective activation of hydrogen peroxide for the degradation of organic dye under visible light irradiation. *Applied Catalysis B: Environmental* 2014; **148–149**: 191–200.
131. Chakraborty A, Islam DA, Acharya H. Facile synthesis of CuO nanoparticles deposited zeolitic imidazolate frameworks (ZIF-8) for efficient photocatalytic dye degradation. *Journal of Solid State Chemistry* 2019; **269**: 566–74.
132. Gao Y, Li S, Li Y, Yao L, Zhang H. Accelerated photocatalytic degradation of organic pollutant over metal-organic framework MIL-53(Fe) under visible LED light mediated by persulfate. *Applied Catalysis B: Environmental* 2017; **202**: 165–74.
133. Ramachandran R, Thangavel S, Minzhang L, Haiquan S, Zong-Xiang X, Wang F. Efficient degradation of organic dye using Ni-MOF derived NiCo-LDH as peroxymonosulfate activator. *Chemosphere* 2020; **271**: 128509.
134. Zhang AY, Wang WK, Pei DN, Yu HQ. Degradation of refractory pollutants under solar light irradiation by a robust and self-protected ZnO/CdS/TiO2 hybrid photocatalyst. *Water Res* 2016; **92**: 78–86.
135. Wang H, Yuan X, Wu Y, Zeng G, Chen X, Leng L, Li H. Synthesis and applications of novel graphitic carbon nitride/metal-organic frameworks mesoporous photocatalyst for dyes removal. *Applied Catalysis B: Environmental* 2015; **174–175**: 445–54.

136. Bandala ER, Peláez MA, Dionysiou DD, Gelover S, Garcia J, Macías D. Degradation of 2,4-dichlorophenoxyacetic acid (2,4-D) using cobalt-peroxymonosulfate in Fenton-like process. *Journal of Photochemistry and Photobiology A: Chemistry* 2007; **186**: 357–63.
137. Shukla P, Fatimah I, Wang S, Ang HM, Tadé MO. Photocatalytic generation of sulphate and hydroxyl radicals using zinc oxide under low-power UV to oxidise phenolic contaminants in wastewater. *Catalysis Today* 2010; **157**: 410–4.
138. Sun H, Liu S, Liu S, Wang S. A comparative study of reduced graphene oxide modified TiO2, ZnO and Ta2O5 in visible light photocatalytic/photochemical oxidation of methylene blue. *Applied Catalysis B: Environmental* 2014; **146**: 162–8.
139. Yang Q, Choi H, Chen Y, Dionysiou DD. Heterogeneous activation of peroxymonosulfate by supported cobalt catalysts for the degradation of 2,4-dichlorophenol in water: The effect of support, cobalt precursor, and UV radiation. *Applied Catalysis B: Environmental* 2008; **77**: 300–7.
140. Quang TT, Truong NX, Minh TH, Tue NN, Ly GTP. Enhanced Photocatalytic Degradation of MB Under Visible Light Using the Modified MIL-53(Fe). *Topics in Catalysis* 2020.
141. Pérez-Lemus N, López-Serna R, Pérez-Elvira SI, Barrado E. Analytical methodologies for the determination of pharmaceuticals and personal care products (PPCPs) in sewage sludge: A critical review. *Analytica Chimica Acta* 2019; **1083**: 19–40.
142. Yu F, Li Y, Han S, Ma J. Adsorptive removal of antibiotics from aqueous solution using carbon materials. *Chemosphere* 2016; **153**: 365–85.
143. Hena S, Gutierrez L, Croué J-P. Removal of pharmaceutical and personal care products (PPCPs) from wastewater using microalgae: A review. *Journal of Hazardous Materials* 2021; **403**: 124041.
144. Jackson CM, Esnouf MP, Winzor DJ, Duewer DL. Defining and measuring biological activity: applying the principles of metrology. *Accreditation and Quality Assurance* 2007; **12**: 283–94.
145. Freyria FS, Geobaldo F, Bonelli B. Nanomaterials for the Abatement of Pharmaceuticals and Personal Care Products from Wastewater. *Applied Sciences* 2018; **8**: 170.
146. Löffler D, Römbke J, Meller M, Ternes TA. Environmental Fate of Pharmaceuticals in Water/Sediment Systems. *Environmental Science & Technology* 2005; **39**: 5209–18.
147. Oberoi AS, Jia Y, Zhang H, Khanal SK, Lu H. Insights into the Fate and Removal of Antibiotics in Engineered Biological Treatment Systems: A Critical Review. *Environmental Science & Technology* 2019; **53**: 7234–64.
148. Bueno MJM, Gomez MJ, Herrera S, Hernando MD, Agüera A, Fernández-Alba AR. Occurrence and persistence of organic emerging contaminants and priority pollutants in five sewage treatment plants of Spain: Two years pilot survey monitoring. *Environmental Pollution* 2012; **164**: 267–73.
149. Guo X, Yan Z, Zhang Y, Kong X, Kong D, Shan Z, Wang N. Removal mechanisms for extremely high-level fluoroquinolone antibiotics in pharmaceutical wastewater treatment plants. *Environ Sci Pollut Res Int* 2017; **24**: 8769–77.
150. Hu Y, Gao X, Yu L, Wang Y, Ning J, Xu S, Lou XW. Carbon-Coated CdS Petalous Nanostructures with Enhanced Photostability and Photocatalytic Activity. *Angewandte Chemie International Edition* 2013; **52**: 5636–9.
151. Kumar A, Khan M, He J, Lo IMC. Recent developments and challenges in practical application of visible–light–driven TiO2–based heterojunctions for PPCP degradation: A critical review. *Water Res* 2020; **170**: 115356.

152. Rasheed T, Bilal M, Hassan AA, Nabeel F, Bharagava RN, Romanholo Ferreira LF, Tran HN, Iqbal HMN. Environmental threatening concern and efficient removal of pharmaceutically active compounds using metal-organic frameworks as adsorbents. *Environmental Research* 2020; **185**: 109436.

153. Dong W, Wang D, Wang H, Li M, Chen F, Jia F, Yang Q, Li X, Yuan X, Gong J, Li H, Ye J. Facile synthesis of In2S3/UiO-66 composite with enhanced adsorption performance and photocatalytic activity for the removal of tetracycline under visible light irradiation. *Journal of Colloid and Interface Science* 2019; **535**: 444–57.

154. Liang R, Luo S, Jing F, Shen L, Qin N, Wu L. A simple strategy for fabrication of Pd@MIL-100(Fe) nanocomposite as a visible-light-driven photocatalyst for the treatment of pharmaceuticals and personal care products (PPCPs). *Applied Catalysis B: Environmental* 2015; 240–248.

155. Wang D, Jia F, Wang H, Chen F, Fang Y, Dong W, Zeng G, Li X, Yang Q, Yuan X. Simultaneously efficient adsorption and photocatalytic degradation of tetracycline by Fe-based MOFs. *Journal of Colloid and Interface Science* 2018; **519**: 273–84.

156. He L, Dong Y, Zheng Y, Jia Q, Shan S, Zhang Y. A novel magnetic MIL-101(Fe)/ TiO2 composite for photo degradation of tetracycline under solar light. *Journal of Hazardous Materials* 2019; **361**: 85–94.

157. Gao Y, Yu G, Liu K, Deng S, Wang B, Huang J, Wang Y. Integrated adsorption and visible-light photodegradation of aqueous clofibric acid and carbamazepine by a Fe-based metal-organic framework. *Chemical Engineering Journal* 2017; **330**: 157–65.

158. Yang C, You X, Cheng J, Zheng H, Chen Y. A novel visible-light-driven In-based MOF/graphene oxide composite photocatalyst with enhanced photocatalytic activity toward the degradation of amoxicillin. *Applied Catalysis B: Environmental* 2017; **200**: 673–80.

159. Wang C, Xue Y, Wang P, Ao Y. Effects of water environmental factors on the photocatalytic degradation of sulfamethoxazole by AgI/UiO-66 composite under visible light irradiation. *Journal of Alloys and Compounds* 2018; **748**: 314–22.

160. Tang Y, Yin X, Mu M, Jiang Y, Li X, Zhang H, Ouyang T. Anatase TiO2@MIL-101(Cr) nanocomposite for photocatalytic degradation of bisphenol A. *Colloids and Surfaces A: Physicochemical and Engineering Aspects* 2020; **596**: 124745.

161. El-Fawal EM, Younis SA, Zaki T. Designing AgFeO2-graphene/Cu2(BTC)3 MOF heterojunction photocatalysts for enhanced treatment of pharmaceutical wastewater under sunlight. *Journal of Photochemistry and Photobiology A: Chemistry* 2020; **401**: 112746.

162. Liang R, Huang R, Ying S, Wang X, Yan G, Wu L. Facile in situ growth of highly dispersed palladium on phosphotungstic-acid-encapsulated MIL-100(Fe) for the degradation of pharmaceuticals and personal care products under visible light. *Nano Research* 2017; 1109–1123.

163. Askari N, Beheshti M, Mowla D, Farhadian M. Fabrication of CuWO4/Bi2S3/ ZIF67 MOF: A novel double Z-scheme ternary heterostructure for boosting visible-light photodegradation of antibiotics. *Chemosphere* 2020; **251**: 126453.

164. Jiang W, Li Z, Liu C, Wang D, Yan G, Liu B, Che G. Enhanced visible-light-induced photocatalytic degradation of tetracycline using BiOI/MIL-125(Ti) composite photocatalyst. *Journal of Alloys and Compounds* 2021; **854**: 157166.

165. Chaturvedi G, Kaur A, Kansal SK. CdS-Decorated MIL-53(Fe) Microrods with Enhanced Visible Light Photocatalytic Performance for the Degradation of Ketorolac Tromethamine and Mechanism Insight. *The Journal of Physical Chemistry C* 2019; **123**: 16857–67.

166. Miao S, Zhang H, Cui S, Yang J. Improved photocatalytic degradation of ketoprofen by Pt/MIL-125(Ti)/Ag with synergetic effect of Pt-MOF and MOF-Ag double interfaces: Mechanism and degradation pathway. *Chemosphere* 2020; **257**: 127123.

167. Huang W, Jing C, Zhang X, Tang M, Tang L, Wu M, Liu N. Integration of plasmonic effect into spindle-shaped MIL-88A(Fe): Steering charge flow for enhanced visible-light photocatalytic degradation of ibuprofen. *Chemical Engineering Journal* 2018; **349**: 603–12.
168. He X, Nguyen V, Jiang Z, Wang D, Zhu Z, Wang W-N. Highly-oriented one-dimensional MOF-semiconductor nanoarrays for efficient photodegradation of antibiotics. *Catalysis Science & Technology* 2018; **8**: 2117–23.
169. Bagheri S, TermehYousefi A, Do T-O. Photocatalytic pathway toward degradation of environmental pharmaceutical pollutants: structure, kinetics and mechanism approach. *Catalysis Science & Technology* 2017; **7**: 4548–69.
170. Feng M, Zhang P, Zhou H-C, Sharma VK. Water-stable metal-organic frameworks for aqueous removal of heavy metals and radionuclides: A review. *Chemosphere* 2018; **209**: 783–800.
171. Fu Z, Guo W, Dang Z, Hu Q, Wu F, Feng C, Zhao X, Meng W, Xing B, Giesy JP. Refocusing on Nonpriority Toxic Metals in the Aquatic Environment in China. *Environmental Science & Technology* 2017; **51**: 3117–8.
172. Joseph L, Jun B-M, Flora JRV, Park CM, Yoon Y. Removal of heavy metals from water sources in the developing world using low-cost materials: A review. *Chemosphere* 2019; **229**: 142–59.
173. Kobielska PA, Howarth AJ, Farha OK, Nayak S. Metal–organic frameworks for heavy metal removal from water. *Coordination Chemistry Reviews* 2018; **358**: 92–107.
174. Ap D, Duwig C, Spadini L, Uzu G, Guedron S, Morel M, Cortez R, Ramos Ramos O, Chincheros J, Martins J. How Uncontrolled Urban Expansion Increases the Contamination of the Titicaca Lake Basin (El Alto, La Paz, Bolivia). *Water, Air, & Soil Pollution* 2017; 1–17.
175. Wu Q, Leung JY, Geng X, Chen S, Huang X, Li H, Huang Z, Zhu L, Chen J, Lu Y. Heavy metal contamination of soil and water in the vicinity of an abandoned e-waste recycling site: implications for dissemination of heavy metals. *Sci Total Environ* 2015; **506–507**: 217–25.
176. Chandra R, Dass S, Tomar P, Tiwari M. Cadmium, carcinogen, co-carcinogen and anti carcinogen. *Indian journal of clinical biochemistry : IJCB* 2001; **16**: 145–52.
177. Bazrafshan E, Mohammadi L, Ansari-Moghaddam A, Mahvi AH. Heavy metals removal from aqueous environments by electrocoagulation process- a systematic review. *Journal of environmental health science & engineering* 2015; **13**: 74-.
178. Kim S, Chu KH, Al-Hamadani YAJ, Park CM, Jang M, Kim D-H, Yu M, Heo J, Yoon Y. Removal of contaminants of emerging concern by membranes in water and wastewater: A review. *Chemical Engineering Journal* 2018; **335**: 896–914.
179. Jiang Y, Liu Z, Zeng G, Liu Y, Shao B, Li Z, Liu Y, Zhang W, He Q. Polyaniline-based adsorbents for removal of hexavalent chromium from aqueous solution: a mini review. *Environmental Science and Pollution Research* 2018; **25**: 6158–74.
180. Li PS, Tao HC. Cell surface engineering of microorganisms towards adsorption of heavy metals. *Crit Rev Microbiol* 2015; **41**: 140–9.
181. Liu X, Pan L, Lv T, Zhu G, Sun Z, Sun C. Microwave-assisted synthesis of CdS–reduced graphene oxide composites for photocatalytic reduction of Cr(vi). *Chemical Communications* 2011; **47**: 11984–6.
182. Yang L, Xiao Y, Liu S, Li Y, Cai Q, Luo S, Zeng G. Photocatalytic reduction of Cr(VI) on WO3 doped long TiO2 nanotube arrays in the presence of citric acid. *Applied Catalysis B: Environmental* 2010; **94**: 142–9.
183. Wang C-C, Zhang Y-Q, Li J, Wang P. Photocatalytic CO2 reduction in metal–organic frameworks: A mini review. *Journal of Molecular Structure* 2015; **1083**: 127–36.

184. Wang C-C, Li J-R, Lv X-L, Zhang Y-Q, Guo G. Photocatalytic organic pollutants degradation in metal–organic frameworks. *Energy & Environmental Science* 2014; **7**: 2831–67.
185. Jing F, Liang R, Xiong J, Chen R, Zhang S, Li Y, Wu L. MIL-68(Fe) as an efficient visible-light-driven photocatalyst for the treatment of a simulated waste-water contain Cr(VI) and Malachite Green. *Applied Catalysis B: Environmental* 2017; **206**: 9–15.
186. Wang L, Wang N, Zhu L, Yu H, Tang H. Photocatalytic reduction of Cr(VI) over different TiO2 photocatalysts and the effects of dissolved organic species. *Journal of Hazardous Materials* 2008; **152**: 93–9.
187. Shi L, Wang T, Zhang H, Chang K, Meng X, Liu H, Ye J. An Amine-Functionalized Iron(III) Metal-Organic Framework as Efficient Visible-Light Photocatalyst for Cr(VI) Reduction. *Adv Sci (Weinh)* 2015; **2**: 1500006.
188. Wang X, Liu J, Leong S, Lin X, Wei J, Kong B, Xu Y, Low ZX, Yao J, Wang H. Rapid Construction of ZnO@ZIF-8 Heterostructures with Size-Selective Photocatalysis Properties. *ACS Appl Mater Interfaces* 2016; **8**: 9080–7.
189. Shen L, Wu W, Liang R, Lin R, Wu L. Highly dispersed palladium nanoparticles anchored on UiO-66(NH2) metal-organic framework as a reusable and dual functional visible-light-driven photocatalyst. *Nanoscale* 2013; **5**: 9374–82.
190. Huang W, Liu N, Zhang X, Wu M, Tang L. Metal organic framework g-C3N4/MIL-53(Fe) heterojunctions with enhanced photocatalytic activity for Cr(VI) reduction under visible light. *Applied Surface Science* 2017; **425**: 107–16.
191. Liang R, Shen L, Jing F, Qin N, Wu L. Preparation of MIL-53(Fe)-Reduced Graphene Oxide Nanocomposites by a Simple Self-Assembly Strategy for Increasing Interfacial Contact: Efficient Visible-Light Photocatalysts. *ACS Applied Materials & Interfaces* 2015; **7**: 9507–15.
192. Liang R, Shen L, Jing F, Wu W, Qin N, Lin R, Wu L. NH2-mediated indium metal–organic framework as a novel visible-light-driven photocatalyst for reduction of the aqueous Cr(VI). *Applied Catalysis B: Environmental* 2015; **162**: 245–51.
193. Wang H, Yuan X, Wu Y, Zeng G, Chen X, Leng L, Wu Z, Jiang L, Li H. Facile synthesis of amino-functionalized titanium metal-organic frameworks and their superior visible-light photocatalytic activity for Cr(VI) reduction. *Journal of Hazardous Materials* 2015; **286**: 187–94.
194. Liu Q, Zeng C, Ai L, Hao Z, Jiang J. Boosting visible light photoreactivity of photoactive metal-organic framework: Designed plasmonic Z-scheme Ag/AgCl@MIL-53-Fe. *Applied Catalysis B: Environmental* 2018; **224**: 38–45.
195. Yadav M, Xu Q. Catalytic chromium reduction using formic acid and metal nanoparticles immobilized in a metal–organic framework. *Chemical Communications* 2013; **49**: 3327–9.
196. Yi X-H, Wang F-X, Du X-D, Fu H, Wang C-C. Highly efficient photocatalytic Cr(VI) reduction and organic pollutants degradation of two new bifunctional 2D Cd/Co-based MOFs. *Polyhedron* 2018; **152**: 216–24.
197. Guan R, Yuan X, Wu Z, Wang H, Jiang L, Li Y, Zeng G. Functionality of surfactants in waste-activated sludge treatment: A review. *Science of The Total Environment* 2017; **609**: 1433–42.
198. Wang C-C, Du X-D, Li J, Guo X-X, Wang P, Zhang J. Heterogeneous photocatalytic reactions of nitrite oxidation and Cr(VI) reduction on iron-doped titania prepared by the wet impregnation method. *Applied Catalysis B: Environmental* 1998; **16**: 187–96.
199. Zhao S, Chen Z, Shen J, Qu Y, Wang B, Wang X. Enhanced Cr(VI) removal based on reduction-coagulation-precipitation by NaBH4 combined with fly ash leachate as a catalyst. *Chemical Engineering Journal* 2017; **322**: 646–56.

200. Wang C-C, Du X-D, Li J, Guo X-X, Wang P, Zhang J. Photocatalytic Cr(VI) reduction in metal-organic frameworks: A mini-review. *Applied Catalysis B: Environmental* 2016; **193**: 198–216.
201. Ighalo JO, Adeniyi AG, Adelodun AA. Recent advances on the adsorption of herbicides and pesticides from polluted waters: performance evaluation via physical attributes. *Journal of Industrial and Engineering Chemistry* 2020. **93**: 117–137.
202. Vaya D, Surolia PK. Semiconductor based photocatalytic degradation of pesticides: An overview. *Environmental Technology & Innovation* 2020; **20**: 101128.
203. Oladipo AA. MIL-53 (Fe)-based photo-sensitive composite for degradation of organochlorinated herbicide and enhanced reduction of Cr(VI). *Process Safety and Environmental Protection* 2018; **116**: 413–23.
204. Oladipo AA, Vaziri R, Abureesh MA. Highly robust AgIO3/MIL-53 (Fe) nanohybrid composites for degradation of organophosphorus pesticides in single and binary systems: Application of artificial neural networks modelling. *Journal of the Taiwan Institute of Chemical Engineers* 2018; **83**: 133–42.
205. Xue Y, Wang P, Wang C, Ao Y. Efficient degradation of atrazine by BiOBr/UiO-66 composite photocatalyst under visible light irradiation: Environmental factors, mechanisms and degradation pathways. *Chemosphere* 2018; **203**: 497–505.
206. Jun LY, Yon LS, Mubarak NM, Bing CH, Pan S, Danquah MK, Abdullah EC, Khalid M. An overview of immobilized enzyme technologies for dye and phenolic removal from wastewater. *Journal of Environmental Chemical Engineering* 2019; **7**: 102961.
207. Zhang C-L, Yu Y-Y, Fang Z, Naraginti S, Zhang Y, Yong Y-C. Recent advances in nitroaromatic pollutants bioreduction by electroactive bacteria. *Process Biochemistry* 2018; **70**: 129–35.
208. Li J, Yang J, Liu YY, Ma JF. Two heterometallic-organic frameworks composed of iron(III)-salen-based ligands and d(10) metals: gas sorption and visible-light photocatalytic degradation of 2-chlorophenol. *Chemistry* 2015; **21**: 4413–21.
209. Masoomi MY, Bagheri M, Morsali A, Junk PC. High photodegradation efficiency of phenol by mixed-metal–organic frameworks. *Inorganic Chemistry Frontiers* 2016; **3**: 944–51.
210. Wu X-Y, Qi H-X, Ning J-J, Wang J-F, Ren Z-G, Lang J-P. One silver(I)/tetraphosphine coordination polymer showing good catalytic performance in the photodegradation of nitroaromatics in aqueous solution. *Applied Catalysis B: Environmental* 2015; **168–169**: 98–104.
211. Li Z-Q, Wang A, Guo C-Y, Tai Y-F, Qiu L-G. One-pot synthesis of metal–organic framework@SiO2 core–shell nanoparticles with enhanced visible-light photoactivity. *Dalton Transactions* 2013; **42**: 13948–54.
212. Cassano AE, Alfano OM. Reaction engineering of suspended solid heterogeneous photocatalytic reactors. *Catalysis Today* 2000; **58**: 167–97.
213. Wang D, Li Z. Iron-based metal–organic frameworks (MOFs) for visible-light-induced photocatalysis. *Research on Chemical Intermediates* 2017; **43**: 5169–86.
214. Shi LE, Li ZH, Zheng W, Zhao YF, Jin YF, Tang ZX. Synthesis, antibacterial activity, antibacterial mechanism and food applications of ZnO nanoparticles: a review. *Food Addit Contam Part A Chem Anal Control Expo Risk Assess* 2014; **31**: 173–86.
215. Lavand AB, Malghe YS. Synthesis, Characterization, and Visible Light Photocatalytic Activity of Nanosized Carbon Doped Zinc Oxide. *International Journal of Photochemistry* 2015; **2015**: 790153.
216. Meng X-B, Sheng J-L, Tang H-L, Sun X-J, Dong H, Zhang F-M. Metal-organic framework as nanoreactors to co-incorporate carbon nanodots and CdS quantum dots into the pores for improved H2 evolution without noble-metal cocatalyst. *Applied Catalysis B: Environmental* 2019; **244**: 340–6.

217. Guo F, Guo J-H, Wang P, Kang Y-S, Liu Y, Zhao J, Sun W-Y. Facet-dependent photocatalytic hydrogen production of metal–organic framework NH2-MIL-125(Ti). *Chemical Science* 2019; **10**: 4834–8.
218. Ahmed S, Rasul MG, Brown R, Hashib MA. Influence of parameters on the heterogeneous photocatalytic degradation of pesticides and phenolic contaminants in wastewater: A short review. *Journal of Environmental Management* 2011; **92**: 311–30.
219. Adesina AA. Industrial exploitation of photocatalysis: progress, perspectives and prospects. *Catalysis Surveys from Asia* 2004; **8**: 265–73.
220. Haque MM, Muneer M, Bahnemann DW. Semiconductor-Mediated Photocatalyzed Degradation of a Herbicide Derivative, Chlorotoluron, in Aqueous Suspensions. *Environmental Science & Technology* 2006; **40**: 4765–70.
221. Saien J, Khezrianjoo S. Degradation of the fungicide carbendazim in aqueous solutions with UV/TiO(2) process: optimization, kinetics and toxicity studies. *J Hazard Mater* 2008; **157**: 269–76.
222. Chong MN, Lei S, Jin B, Saint C, Chow CWK. Optimisation of an annular photoreactor process for degradation of Congo Red using a newly synthesized titania impregnated kaolinite nano-photocatalyst. *Separation and Purification Technology* 2009; **67**: 355–63.
223. Priya MH, Madras G. Photocatalytic degradation of nitrobenzenes with combustion synthesized nano-TiO2. *Journal of Photochemistry and Photobiology A: Chemistry* 2006; **178**: 1–7.

doi: 10.2166/9781789061932_0073

Chapter 3
Photocatalytic reactor types and configurations

Meysam Shaghaghi[1], Hamed Sargazi[2], Alireza Bazargan[2*] and Marianna Bellardita[3]

[1]School of Chemical Engineering, College of Engineering, University of Tehran, Tehran, Iran
[2]School of Environment, College of Engineering, University of Tehran, Tehran, Iran
[3]Dipartimento di Ingegneria, Università di Palermo, Palermo, Viale delle Scienze 90128, Italy
*Corresponding author: alireza.bazargan@ut.ac.ir

ABSTRACT

The photocatalytic treatment process is generally affected by three main components: the pollutants, the catalysts, and the light source generating the photons. Among these, the pollutants and catalysts are in the liquid and solid phases, respectively, and the photons are in the form of massless particles. The differences in these facets have made the design of photocatalytic reactors a challenging and not-so-straightforward task. As of yet, different reactors in different scales and shapes, ranging from bench-top and laboratory designs to those of semi-industrial and industrial dimensions have been built. The basic parameters that vary in photocatalytic reactor design are light (light wavelength and light source), catalyst form (suspended or immobilized), and the type of reactor (continuous or batch). This chapter examines each of these parameters and their related issues. Different configurations of photocatalytic reactors including fluidized bed, cascade, annular, spinning disk, Taylor vortex, and optical fiber, amongst others will be presented, along with their specifications and usage.

3.1 INTRODUCTION

One of the many applications of photocatalytic processes is water and wastewater treatment. Photocatalytic degradation of pollutions is biocompatible due to its ability to completely mineralize organic pollutants (under certain conditions) and also the lack of a waste stream at the end of the process. As discussed thoroughly in the literature, photons activate photocatalytic particles to produce hydroxyl radicals, which oxidize and convert contaminants such as resistant organic matter, pesticides, herbicides, detergents, pathogens, viruses, coliforms, and so on. into less harmful (and often harmless) end products. Photocatalytic treatment processes are performed using a photocatalytic reactor

© 2022 The Editors. This is an Open Access book chapter distributed under the terms of the Creative Commons Attribution Licence (CC BY-NC-ND 4.0), which permits copying and redistribution for noncommercial purposes with no derivatives, provided the original work is properly cited (https://creativecommons.org/licenses/by-nc-nd/4.0/). This does not affect the rights licensed or assigned from any third party in this book. The chapter is from the book Photocatalytic Water and Wastewater Treatment, Alireza Barzagan (Ed.).

Photocatalytic Water and Wastewater Treatment

or photoreactor. In general, photoreactors are devices that bring the needed light and reactants together in effective and proper contact with each other [1].

An important obstacle in the development of efficient photocatalytic reactors is the design and fabrication of large-scale photoreactors for industrial and commercial use. To achieve this goal, several parameters must be optimized in the design of the reactor, including the following [2]:

- selection of a radiation source according to important parameters such as output power, source efficiency, spectral distribution, shape, dimensions, and operation and maintenance (O&M) requirements;
- design of tools related to the radiation system such as mirrors and reflectors according to the parameters of shape, dimensions, materials used, and their cleaning process;
- selection of a suitable photocatalyst;
- design of the reactor geometry according to the radiation source.

Regarding the light source, there is a need to use lamps whose emission spectrum does not deviate much from lamps used in laboratory photoreactors and bench-scale testing, but which have higher electrical power, longer life, and also more compatibility with the geometry of the photoreactor [3–5].

Sunlight can also be used as a light source in photocatalytic reactors since about 5% of the wavelengths of the solar spectrum (which fall within the ultraviolet (UV) region) can activate TiO_2 photocatalytic particles (for unmodified TiO_2), although the intensity of the light is not constant over time. The relationship between photocatalytic removal rate and light intensity at low intensities is, generally, first order, while at higher levels of radiation, this relationship follows fractional orders [1].

Evaluation of radiation and its distribution inside photocatalytic reactors is necessary to generalize laboratory-scale results to large-scale operations and compare efficiency. In order to properly scale photocatalytic reactors, it is imperative to maximize the number of photons which are adsorbed in the given volume and time. In addition, the electron-holes that are created during the process should not be wasted.

In the case of photocatalytic (nano)particles, choosing the right photocatalyst can pose particular challenges. In addition to conventional TiO_2 (including commercial and synthesized grades in different crystalline phases of anatase, rutile, and brookite), other options such as ZnO, CdS, WO_3, and SnO_2 can be used as photocatalytic particles [1].

For the photocatalytic method to work effectively, mass and photon transfer limitations must be considered in the reactor configuration. Mass and photon transfer characteristics in immobilized and slurry photocatalytic reactors are different. Photocatalytic particles in the slurry state are more active than in the immobilized state, but the need to separate the photocatalytic particles from the output stream of the slurry reactors is an important challenge. Immobilized photocatalytic reactors will perform similarly to slurry reactors if the photocatalyst band gap energy parameters, catalyst thickness, and proximity between the reactants and the catalyst surface, are properly designed to overcome mass and photon transfer limitations [6].

Photocatalytic reactor types and configurations   75

The rate at which pollutants are removed and also the final removal percentage are dependent on various factors. Although the specifications of the catalyst (such as surface, particle size, concentration, etc.) are important, other parameters such as the concentration of the pollutants, pH, temperature, and light wavelength and intensity should not be overlooked [2].

On a micro-level, the physico-chemical principles of photocatalytic systems are important, but on a larger level, reactor design and engineering require deep scrutiny. This is especially true for larger-scale reactors that treat larger volumes of water or wastewater and use more energy. In the following, we first examine the basic components in the design of photocatalytic reactors and then describe some different variations that have been used.

3.2 BASIC COMPONENTS IN THE DESIGN OF PHOTOCATALYTIC REACTORS

Photocatalytic reactors can be classified based on three components: light source, catalyst, and reactor, which are described as follows.

3.2.1 Characteristics of light

Parameters related to the light are the wavelength of the light to activate the catalyst (such as UV or visible radiation) and the light source itself.

3.2.1.1 The wavelength of light

The wavelength threshold for activating a photocatalyst depends on its band gap energy. The value of this threshold is obtained using Planck's equation:

$$\lambda\ (nm) = \frac{1240}{E_{bg}\ (eV)} \tag{3.1}$$

where λ is the wavelength of light and E_{bg} is the band gap energy [7]. For example, the ideal wavelength for a TiO_2 catalyst with a band gap energy equivalent to 3.1 eV is 400 nm [8]. The band gap energy values and wavelengths of some other types of catalysts used in the design of photocatalytic reactors are presented in Table 3.1.

However, by modifying the catalyst, the band gap energy can be changed, resulting in a wavelength threshold shift for catalyst activation. Kumordzi *et al.* (2016) synthesized TiO_2 catalyst modified with graphene (G) and used the TiO_2-G composite catalyst for photocatalytic degradation of Zn^{2+}. The results showed that the band gap energy decreased from 3.1 eV for pure TiO_2 to 2.2 eV for TiO_2-G. In tandem, the value of the wavelength threshold increased from 400 to 563 nm [7].

3.2.1.2 Light source

Light source is one of the most important components in photocatalytic reactors. The light source used in photocatalytic reactors can be natural light (sunlight) or artificial light (by lamps).

76 Photocatalytic Water and Wastewater Treatment

Table 3.1 Band gap energies and corresponding radiation wavelengths for activation of some semiconductors [9].

Semiconductor	Band Gap Energy (eV)	Wavelength (nm)
TiO_2 (Anatase)	3.2	388
TiO_2 (Rutile)	3.03	413
TiO_2 (Brookite)	3.30	376
ZnO	3.2	388
CdS	2.4	516
WO_3	2.8	443
SnO_2	3.6	345

Solar radiation is cheap and depending on the location, is more abundant in terms of UV photon supply. However, radiation flux is highly variable and only available during the day [11]. Photocatalytic reactors with natural light source (solar photoreactors) are divided into two categories: concentrating and non-concentrating light, based on the received radiation. The advantages and disadvantages of these two categories are summarized in Table 3.2.

Verma *et al.* (2018) investigated and compared the photocatalytic decomposition of isoproturon herbicide using a fixed bed solar reactor with a concentrating and non-concentrating system. The results showed that the reactor with the concentrating system increased the removal efficiency, due to the increase in the number of photons to activate the catalyst, and also reduced the decomposition reaction time. The results of this research are presented in Table 3.3 [12] to highlight the superior performance of concentrating systems.

Barzegar *et al.* (2021) investigated the removal of methylene blue (MB) and rhodamine B (RhB) contaminants using a solar parabolic trough photoreactor and g-C_3N_4/TiO_2 photocatalyst. The set-up is shown in Figure 3.1. Removal

Table 3.2 The main advantages and disadvantages of concentrating and non-concentrating solar photocatalytic reactors [2].

Non-Concentrating Reactors	Concentrating Reactors
Advantages	
1. Negligible light losses	1. Smaller reactor volume
2. Use of both direct and diffused light	2. Smaller harvesting area
3. Simple design and low investment cost	3. Possibility to reduce reaction time and increase removal efficiency of pollutants
4. Linear relationship between efficiency and radiation intensity	
Disadvantages	
1. High frictional pressure drop	1. More investment costs
2. Larger reactor	2. Use of only direct light
	3. Square-root dependence between light intensity and efficiency

Table 3.3 A comparative example of using concentrating and non-concentrating solar photocatalytic systems [12].

Reactor	Parameter			
	Removal Efficiency (%)	Reaction Time (minutes)	The Volume of the Treated Solution (L)	Collision Photon Flux (Einstein/L/s)
Fixed bed with concentrating system	91	240	6	8.5×10^{-6}
Fixed bed with non-concentrating system	65	600	6	3.51×10^{-6}

efficiencies for contaminants MB and RhB were 94.94% and 93.07%, respectively [13].

As mentioned, the main disadvantage of solar photoreactors is their variable radiation flux and slow kinetics. Thus, the design of photocatalytic reactors has predominantly employed artificial lighting systems which use UV radiation. UV radiation is divided into three categories according to the electromagnetic

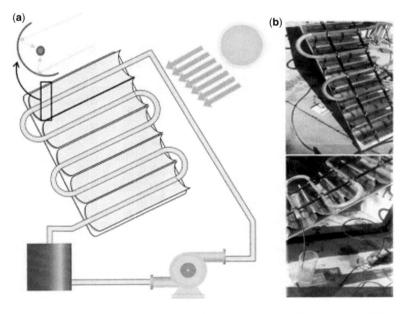

Figure 3.1 The experimental set-up of a solar photoreactor: (a) scheme, and (b) photo. Reproduced from [13].

78 **Photocatalytic Water and Wastewater Treatment**

spectrum: UV-A, UV-B and UV-C. UV-A has a radiation wavelength of 320 to 400 nm, UV-B has a radiation wavelength of 290 to 320 nm and UV-C has a radiation wavelength of 200 to 290 nm [14].

The most common types of UV radiation sources in photocatalytic processes are low- and medium-pressure mercury lamps. Despite having the advantage of constant photon flux, these lamps have disadvantages such as short life (9000–12,000 h), low energy efficiency, and destructive environmental effects due to the presence of mercury [15]. In order to overcome these shortcomings, in some studies high-pressure mercury lamps have been used as the light source [9]. Another type of artificial light source that can be a good alternative to conventional mercury lamps is UV light-emitting diodes (UV-LEDs). The advantages of this light source include low space occupation, durability, faster start-up, lower energy consumption, and longer life (35,000–50,000 h) [16]. Some examples of studies which have used artificial light sources are listed in Table 3.4.

The position of the lamp or radiation source is also a distinguishing feature of photocatalytic reactors. The configuration types of photocatalytic reactors can be divided into three general categories according to the position of the light source [2]:

- Reactors with immersed radiation source (lamp is placed inside the reactor)
- Reactors with external radiation source (lamp is placed outside the reactor)
- Reactors with distributed light sources (radiation is transmitted from the source to the reactor by optical instruments such as reflectors)

Table 3.4 Examples of photocatalytic studies using artificial light.

Photocatalyst	Pollutant	Initial Concentration of Pollutant	Light Source	Removal Efficiency (%)	Reference
ZnO/TiO_2	4-Chlorophenol	25 mg/L	16 W UV-B lamp	About 100	[17]
TiO_2	COD of landfill leachate	28–34 g/L	15 W UV-C lamp	92	[18]
ZnO	4-Chlorophenol, phenol	10 mM	125 W HP Hg lamp	92	[17]
ZnO/SnO_2	MB	10 mg/L	500 W Hg lamp	97.3	[19]
TiO_2/ZnO	Phenol	100 mg/L	UV light	~100	[20]
$ZnO/TiO_2/Au$	Anthracene	50 µg/L	250 W HP Hg lamp	53.7	[17]
$TiS_2–TiO_2$	Acid black 1	200 mg/L	UV light	86	[21]
TiO_2	Imidacloprid	10 mg/L	15 W T5	97	[22]
Mo_2C	Maxilon blue GRL 300 basic dye	–	Tungsten light	90.5	[23]

COD, chemical oxygen demand.

Photocatalytic reactor types and configurations

3.2.2 Catalyst

Several catalysts such as TiO_2, ZnO, WO_3, ZnS, Fe_2O_3, CdS, and so on. have been used for the photocatalytic degradation of various pollutants [24, 25]. One of the basic aspects of designing photocatalytic reactors is how to use the catalyst. In general, in photoreactors, the catalyst can be used as a slurry or immobilized. In the following, each of these two modes will be presented.

3.2.2.1 Slurry photoreactor

Slurry photocatalytic reactors have been used since the early 1990s [26]. In these photoreactors, the catalyst particles are freely dispersed in the liquid phase. Usually, mechanical, magnetic, or gas stirrers and agitators are used to maintain the suspension state of the catalyst in the liquid phase. Among the advantages of using a catalyst as a slurry are the high surface-to-volume ratio due to its relatively uniform distribution, proper mixing of catalyst particles, and easy reactivation of the photocatalyst. However, a continuous filtration process is required to separate the catalytic particles from the treated stream. This separation is usually achieved by a physical process such as filtration or centrifugation, not to mention separation using coagulation and settling, which increases the operating cost of this type of reactor [27]. Another disadvantage of this type of reactor is the possibility of light being scattered by the slurry particles which in turn prevents the penetration of light to all required spaces [28].

In general, the advantages and disadvantages of slurry photoreactors are as follows [2]:

Advantages:

- Relatively uniform distribution of catalysts
- High ratio of catalyst surface to reactor volume
- Simple design and construction of the reactor
- Low-pressure drop in the reactor
- Good mixing of catalyst suspension
- Minimizing the possibility of catalyst clogging due to the possibility of removing or replacing the catalyst continuously during the process

Disadvantages:

- Need to separate photocatalytic particles from the treated stream
- Light scattering by suspended particles

Some types of slurry catalyst photoreactor configurations are the slurry annular reactor (SAR), open upflow reactor (OUR), integrated flow reactor-membrane filtration (IFR-MF), swirl flow reactor (SFR), Taylor vortex reactor (TVR), and turbulent slurry reactor (TSR) [2].

To achieve the optimal design of slurry photocatalytic reactors and subsequently high efficiency in treatment operations, several studies have examined the operating parameters affecting the performance of the reactor. For example, Sivagami *et al.* (2016) investigated the effect of initial concentration, solution pH, and catalyst concentration on the removal of Endosulfan (ES) and chlorpyrifos contaminants (CPS) using a slurry annular photoreactor with a 254 nm UV lamp as a light source. Effect of parameters like concentration

80 Photocatalytic Water and Wastewater Treatment

(5–25 mg/L), pH (3.5 to 10.5) and the catalyst loading (0.5–2 g/L) were investigated on the removal of each pollutant. The removal efficiencies of ES and CPS were reported at 80–99% and 84–94%, respectively [29].

Kamble *et al.* (2006) investigated the effect of solar and UV radiation on the removal process of 1,3-dinitrobenzene (m-DNB) using a slurry photoreactor with TiO_2 photocatalytic particles. The effects of various parameters, such as initial pollutant concentration, catalyst dosage, presence of anions, and pH, were investigated. The results showed the destructive effect of the presence of anions on the rate of photocatalytic degradation of the pollutant. Also, the rate of degradation of the pollutant was maximum at neutral pH [30].

In another study, Nishio *et al.* [31] investigated the decolorization rate of orange II azo dye using a slurry photoreactor containing ZnO photocatalytic nanoparticles under UV irradiation. The material of the cylindrical reactor used in this research was Pyrex glass and its inner diameter, height and volume were 0.08 m, 0.55 m, and 2 liters, respectively. In the design of this reactor, three 15 W UV lamps were used outside the reactor, each placed 25 mm away from the surface of the photoreactor. The effect of various factors such as radiation intensity, initial dye concentration, nanoparticles dosage, and the pH of the solution were investigated. The results showed that with decreasing the initial dye concentration and increasing the intensity of UV radiation, the dye removal efficiency increased and that the highest removal efficiency was obtained at pH = 7.7 [31].

In a study conducted by Wu *et al.* [32], the removal of organic dye from textile industry effluent was investigated using a slurry reactor. The catalyst used in this study was TiO_2 in the form of a slurry, and organic dye and air were injected into the reactor. To investigate the removal kinetics of the contaminant, the effect of mixing speed (50–200 rpm), TiO_2 suspension concentration (0.25–1.71 g/L), initial contaminant concentration (10–50 ppm), temperature (10–50°C), and UV source power intensity (0–96 W) were studied using the one-factor-at-a-time method. The results showed that changing the mixing speed in the mentioned range had little effect on the contaminant removal rate. Also, the maximum contaminant removal rate occurred at the initial concentration of TiO_2 suspension equal to 0.98 g/L. The dye's removal rate increased to a certain value with increasing dye concentration depending on the temperature and then decreased. The pollutant removal rate was maximized when UV power equaled 64 W [32].

In another study, Tokumura *et al.* [33] investigated the removal of azo dye orange II during a homogeneous photocatalytic process (photo-Fenton) using a cylindrical column slurry photoreactor. The iron ions required for this process were obtained by washing tourmaline powder containing 4.49% w/w of Fe_2O_3. The radiation source in this system was placed outside the photoreactor. The results showed that with increasing the amount of tourmaline and UV light intensity as well as decreasing the initial dye concentration, the dye removal efficiency improved [33].

3.2.2.2 Immobilized photoreactors
Immobilized photocatalytic reactors are reactors in which photocatalytic particles are fixed by physical surface forces or chemical bonds on supports.

Photocatalytic reactor types and configurations

Among the characteristics of a good support for photocatalytic particles are the following [34]:

- Transparency against UV rays
- No negative effect of physical surface forces or chemical bonds on the activity of nanoparticles
- Having a high specific surface area
- Having a good adsorption capacity to absorb organic matter and further allow its degradation
- Positive effect on facilitating mass transfer processes
- Chemical inertness
- Ability to recover and reuse

Some supports which have been used in this type of reactor are:

- Carbon fiber [35]
- Stainless steel [36]
- Silica gel [37]
- Activated carbon [38]
- Zeolite [39]
- Glass beads [40]
- Alumina [41]
- Polymer films [42]
- Sand [43]
- Quartz [34]
- Perlite [44]

One of the most important advantages of these reactors is that there is no need to retrieve the catalyst from the liquid stream. This can help the treatment operation proceed continuously with fewer problems.

In general, the advantages and disadvantages of these photoreactors are [2]: Advantages:

- They can be used in continuous processes
- No need for additional processes, such as filtration, to separate the catalyst
- Good efficiency at removing organic pollutants if using supporting materials with high adsorption properties

Disadvantages:

- Low ratio of photocatalyst active surface to reactor volume
- Low light efficiency due to light scattering in the surrounding environment of the photocatalyst
- High-pressure drop if the liquid stream is forced through the support
- Chance of the catalyst washing out from the bed
- Restriction of mass transfer

Types of catalyst immobilized photoreactors include falling film reactor (FFR), multiple tube reactor (MTR), packed bed photoreactor (PBR), fiber optic cable

82 **Photocatalytic Water and Wastewater Treatment**

reactor (FOCR), rotating disk reactor with controlled periodic illumination (RDR-CPI), spiral glass tube reactor (SGTR), and tube light reactor (TLR) [2].

Grieken *et al.* [45] investigated and compared the efficiency of slurry and immobilized photoreactors with the aim of inactivating Escherichia coli. They considered various parameters such as the concentration of TiO_2 nanoparticles in the slurry reactor, and the thickness of the TiO_2 layer in the immobilized reactor. The configuration of the immobilized reactor was designed in two forms: TiO_2 stabilization on the wall and TiO_2 stabilization on the fixed bed packing. According to the obtained results, slurry reactors showed higher efficiency in the inactivation of the microorganisms.

Li *et al.* [46] designed a double-cylindrical-shell (DCS) photoreactor and fixed TiO_2-coated silica gel beads on the outer surface of the inner cylinder. The reactor shell was built from quartz glass. The set-up was used to remove rhodamine B and methyl orange from water and operational parameters, such as flow rate and initial concentration, were investigated. The results showed that this reactor had a higher efficiency in removing the pollutants and used less energy compared to thin-film and slurry photoreactors.

Behnajady *et al.* [47] used a continuous flow photoreactor to degrade C.I. acid red 27 (AR27). The reactor was made of four quartz tubes that were connected in series by a polyethylene tube. TiO_2 photocatalyst particles were loaded onto plates and then placed inside each of the quartz tubes. The light source used was UV lamps placed in front of the tubes. The results showed that with increasing light intensity, the removal efficiency of the desired pollutant increased linearly and decreased with increasing flow rate.

Malakootian *et al.* [48] investigated the photocatalytic degradation of the antibiotic ciprofloxacin using TiO_2 nanoparticles (1 g/L) fixed on a glass plate. The maximum removal efficiency of the mentioned contaminant was obtained under operating conditions of pH = 5, contact time equal to 105 minutes, and initial concentration of the contaminant equal to 3 mg/L. The optimal removal efficiencies of ciprofloxacin in synthesized and real cases were 92.81% and 86.57%, respectively.

The photocatalytic removal of phenol using ZnO nanosheets fixed on montmorillonite was investigated in 2015 by Ye *et al.* [49]. The results showed that if ZnO nanosheets immobilized on montmorillonite were used, the removal efficiency increased by about 32.5% compared to the use of ZnO photocatalytic particles (88.5% vs. 56%).

Damodar *et al.* [50] used an innovative photoreactor with TiO_2 nanoparticles on polyvinyl chloride (PVC) rotating tubes to degrade azo dye. In this reactor, the fluid flow was continuous and three 30 W low-pressure mercury UV lamps were used to emit light. The reactor consisted of two cells, each containing five tubes of TiO_2-coated PVC. Both cells had the same volume (1.5 L) and the same light distribution. All tubes were rotated simultaneously by a variable speed DC motor. In this study, the effect of key parameters of initial dye concentration, tube rotation speed, pH, and fluid flowrate on the removal of the desired contaminant was investigated. The results showed that low initial contaminant concentrations and acidic conditions are suitable for pollutant

Photocatalytic reactor types and configurations 83

removal. The optimum speed of the PVC pipes was 25 rpm. Dye removal efficiency of 90–99.9% and total organic carbon (TOC) removal efficiency of 55–70% were obtained.

3.2.3 Process type
Processes can be divided into batch and continuous configuration, each employing specific types of reactors as described below.

3.2.3.1 Batch reactors
In batch reactors, the reactants enter the reactor and react with each other over a period of time. At the end of the reaction, the products and unused reactants (if any) are removed from the reactor [51]. This means that these types of reactors have no inlet or outlet flow of reactants or products during the reaction. In this type of process, mixing operations are often used to maintain the uniformity of the reaction mixture, and if the reaction mixture is completely stirred, the reaction rate will not vary depending on the position within the reactor. In other words, in an ideally mixed batch reactor, the reaction mixture's properties and the reaction rate are the same at all points in the reactor.

Batch reactors are quick and easy to install, and also easy to operate, yet they may require a continuous flow of cold water for cooling due to the increase of temperature resulting from the photocatalytic lamp [1, 52, 53].

Numerous studies have been performed on the removal of various contaminants using slurry or immobilized photocatalytic particles in batch photoreactors. For example, Divya et al. [54] studied the photocatalytic degradation of an acidic dye, orange G, using a UV irradiated batch photoreactor with H_2O_2 injection. Various process parameters such as pH, dye concentration, amount of H_2O_2 and TiO_2, light source, and light intensity along with the effect of reflective surfaces were examined. In this study, a reactor with dimensions of $56 \times 28.5 \times 59$ cm with two 15 W UV lamps and a high-pressure 125 W lamp were used. The results showed that if the reflecting surface is used, a higher removal efficiency, compared to when the surfaces are painted black (100% vs. 85%), is obtained.

Chanathaworn et al. [55] used a batch photoreactor containing 0.5 L of solution to remove rhodamine B (RhB) and malachite green (MG). The photoreactor was placed inside a stainless steel chamber and UV lamps were mounted vertically on the inner wall of the chamber. The reactor was cooled by airflow. The effects of parameters of initial pollutant concentration (10–30 mg/L), light intensity (0–114 W/m^2) and TiO_2 concentration (0.5–1.5 g/L) were investigated. Due to the structure of the MG contaminant, its removal rate was higher than that of RhB. According to the experiments, the color removal efficiency increased with increasing the amount of TiO_2 and the intensity of UV light. However, with increasing initial pollutant concentration, this efficiency decreased. The maximum dye removal efficiency was reported at 1.5 g/L catalyst concentration, 20 mg/L initial pollutant concentration, and 114 W/m^2 light intensity. A diagram of the reactor used in this study is presented in Figure 3.2 [55].

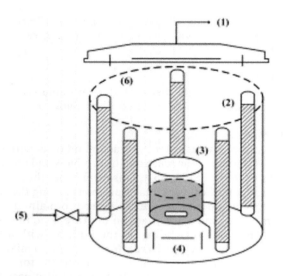

Figure 3.2 Laboratory set-up of a batch reactor used for rhodamine B (RhB) and malachite green (MG) removal: (1) cooling air outlet, (2) UV lamps, (3) reactor, (4) stirrer, (5) cooling air inlet, (6) chamber. Reproduced from [55].

3.2.3.2 Continuous reactors

Continuous photoreactors are classified into continuous stirred tank reactors (CSTRs) and plug flow reactors (PFRs) based on their hydrodynamic regime. Studies on PFR and CSTR systems are imperative if photocatalysis is to be used in large-scale applications [56].

As the name implies, a CSTR has an inlet and outlet flow, and the contents are constantly stirred. Therefore, due to complete mixing, the concentration of components and temperature, and consequently the reaction rate, will be the same in all parts of the reactor. In this case, the concentration and temperature of the output of the reactor are equal to the concentration and temperature of the contents inside the reactor. It should be noted that at the beginning of the operation of these reactors, the reaction conditions are transitional and unstable before stability is reached after a short amount of time.

Behnajady *et al.* [57] designed a batch-recirculated photoreactor to remove C.I. acid red 17. For this purpose, an annular photoreactor was combined with a CSTR and the output flow from the CSTR was returned to the annular reactor. The effect of operating parameters such as fluid volume inside the CSTR and volumetric flow rate was investigated. The results showed that the removal efficiency of the mentioned pollutant increased with increasing volumetric flow rate, while it decreased with increasing the volume of the liquid in the CSTR. In another study, Sheidaei and Behnajady [58] used TiO_2-P25 nanoparticles in the reactor mentioned in the previous study for the removal of acid orange 7. They examined various parameters such as initial pollutant concentration, solution volume, volumetric flow, reaction time, and light source power. The experimental results demonstrated that the key parameter affecting the efficiency of pollutant removal was the light source. Optimal values for initial pollutant concentration, solution volume, volume flow, reaction time, and light source power were found to be 20 mg/L, 500 mL, 140 mL/min, 60 min, and 13 W, respectively.

On the other hand, in plug flow reactors (PFRs) the reaction solution enters from one side and exits from the other side, but follows a straight path within a tube without any mixers. Ideally, the fluid flow is assumed to be turbulent (Reynolds number greater than 2000) resulting in fluid motion along the bed as a piston with no axial dispersion. In this case, the reactants flow continuously along the reactor, and their consumption and concentration are constantly changing, but the concentration changes in the direction of the reactor radius become negligible. Of course, this is not always the case. For example, Chen and et al. [59] used a simple tubular reactor to remove organic dye from the environment. The Reynolds number obtained for the volumetric velocity inside the reactor was 184, indicating a laminar flow inside the reactor.

Alam *et al.* [60] investigated the performance of continuous and batch reactors in the photocatalytic removal of reactive yellow dye using TiO_2 catalyst immobilized on ceramic plates. The operating parameters including initial dye concentration, amount of catalyst and reaction time were studied. The results showed that the batch reactor had a higher removal efficiency and by using this type of reactor, under a residence time of 360 minutes, the removal efficiency of the reactive yellow dye with an initial concentration of 200 ppm was 60%.

3.3 PHOTOREACTOR CONFIGURATIONS

The configuration of the reactor and its operating conditions can strongly affect the performance and extent of the photocatalytic reactions. There have been numerous reactors used at different scales (from bench-top to industrial scale) for photocatalysis. The shapes and scales of reactors used is diverse, and the following have been chosen as some of the most common types to be further discussed:

- Fixed bed photoreactors
- Packed bed photoreactors (PBR)
- Fluidized bed reactors
- Thin film reactors
- Membrane reactors
- Cascade reactors
- Spinning disk reactors
- Rotating photocatalytic reactors
- Taylor vortex reactors
- Optical fiber reactors

3.3.1 Bed reactors

In the last decade, photocatalyst stabilization on the substrate (based on physical surface forces or chemical bonding) has received much attention. In fixed bed reactors, the contact of the contaminant with the catalyst surface and the amount of mass transfer is less than with a suspended bed, and the performance of this type of photoreactor depends to a large extent on the surface area of the bed [61]. Various forms of these photoreactors have been studied to overcome the limitation of mass transfer. In these studies, flow mixing is often created by stirring velocity, rotational velocity, cascading flow, or reactor geometry.

3.3.1.1 Fixed bed photoreactor

Alexiadis and Mazzarino used a fixed bed photoreactor for wastewater treatment at pilot and industrial scale [62]. The cost of wastewater treatment was determined by taking into account the energy consumption and periodic need for replacing commercial UV lamps. It was found that the optimal conditions for wastewater treatment depend on the degradation kinetics of the pollutants. Using high-power UV lamps and dense catalysts, the process cost was reduced and the degradation reaction rate was slow. Conversely, a low power UV light source and catalyst with low absorption were used to achieve a high rate of degradation [62].

Cloteaux *et al.* [63] studied the removal of formaldehyde using a fixed bed photocatalytic reactor filled with TiO_2-coated media as shown in Figure 3.3. The media used was Raschig rings and the light source was UV-A lamps. The description of the hydraulic behavior of the reactor in this study is based on the residence time distribution model, which includes light distribution, chemical kinetics and mass transfer. This model, along with the Langmuir–Hinshelwood

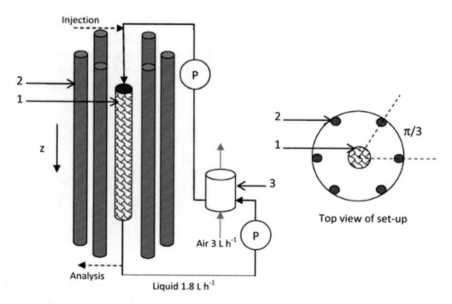

Figure 3.3 Experimental fixed bed photocatalytic reactor filled with TiO_2-coated media: (1) fixed bed reactor (2) lamps and (3) tank. Reproduced from [63].

(L-H) kinetic model, determined the changes in the concentration of the pollutant at the reactor output. The favorable results showed high formaldehyde removal efficiency.

Packed bed reactors usually have an annular geometry so that the light source is located on the central axis. However, this configuration limits photocatalytic radiation and non-uniform flow distribution, and this limitation has a negative effect on the overall efficiency of the photoreactor. Compared to annular photoreactors, reactors with flat plate geometry can be used at more diverse scales and can also be activated by sunlight if need be. In general, the flow regime of the liquid must be selected in such a way that adequate mixing is achieved, while the surface of the photocatalyst particles is not damaged by high velocities, perturbations, and shear stresses [64].

Vaiano *et al.* [65] immobilized nitrogen-doped TiO_2 particles on glass spheres and used them in a fixed bed reactor with a flat plate geometry to maximize contact between photocatalyst particles and radiation for wastewater treatment. The results showed that the removal of methylene blue increases with increasing contact time, whether the light source is UV or visible light.

3.3.1.2 *Packed bed photoreactor (PBR)*
Borges *et al.* [66] used a continuous PBR reactor to remove paracetamol contaminant from wastewater. In the study two configurations were tested, namely, (a) a CSTR photoreactor with TiO_2 suspended particles, and (b) a PBR reactor with TiO_2 immobilized on glass balls as depicted in Figure 3.4. To achieve optimal conditions, the amount of titanium, pH, and wastewater concentration were investigated. The results of this study indicated that the PBR reactor showed better efficiency in removing the desired contaminant.

3.3.2 Fluidized bed reactors
One of the most commonly used catalytic reactors is the fluidized bed design. Although these reactors are heterogeneous, they have good mixing, so the temperature and concentration are evenly distributed throughout the bed. This type of reactor contains a large amount of feed as well as solids and has good temperature control. Slurry reactors, although highly effective, face major limitations in industrialization due to the problem of catalyst particle separation, especially when very fine particles are used. In this type of reactor, the catalyst can be fixed on the fluidized substrate as well. When a fluid stream does not pass through the system, the particles settle to the bottom of the reactor and the system remains idle. Only when fluid passes through the particles (from the bottom towards the top) do the particles jump, mix and become fluidized as the name of the system implies. In photocatalytic processes, photocatalytic particles are usually immobilized on substrates such as glass pellets, ceramic particles, and activated carbon. In this regard, Mohammad Kabir and his colleagues immobilized TiO_2 photocatalytic particles on a glass substrate and studied the removal of pollution from the aqueous medium and the effect of adding hydrogen peroxide using a fluidized bed reactor as shown in Figure 3.5 [67].

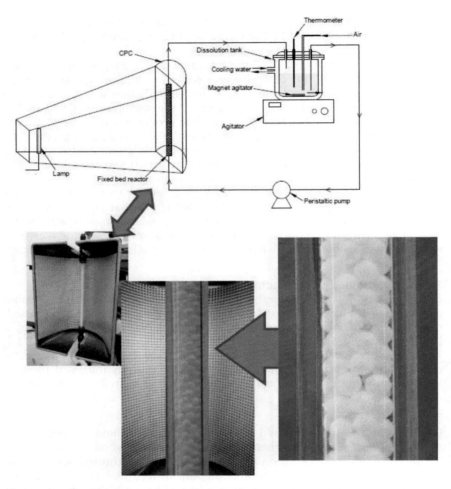

Figure 3.4 Packed bed photoreactor with glass balls as the bed (CPC: compound parabolic collector). Reproduced from [66].

3.3.3 Thin-film reactors

Thin-film reactors are designed to perform best with low reactive volumes (50 to 500 mL) and concentrated solutions in which radiation penetrates millimeters. The reaction solution is injected into the reactor through a glass jet from the tank, and a thin film of fluid flows by gravity onto a closed transparent tube. A low-pressure mercury or phosphor-coated lamp is placed inside the tube, which provides uniform light radiation for the thin film. For the reverse flow direction mode, the direction of fluid flow is from bottom to top.

Quartz or borosilicate double-walled glass tubes are often used in this type of reactor. The inlet flow rate of the liquid jet varies from 0.1 to 1 mL/s, depending

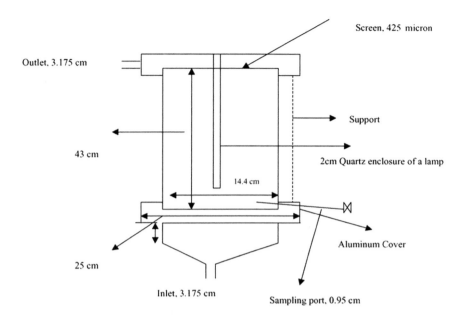

Figure 3.5 Schematic of a fluidized bed reactor used for experimentation. Reproduced from [67].

on the size of the jet and the setting of the bypass flow control. The liquid film thickness is in the range of 0.1 to 0.3 mm depending on the flow rate and solvent used. Self-priming pumps are used in these reactors to inject a mixture of air and fluid. Thin-film reactors work best to control changes in a photochemical reaction during the discharge of the reactant or the formation of a product.

Damodar et al. [68] used a thin-film plate reactor immobilized with TiO_2 nanoparticles under sunlight to investigate the removal of four dye contaminants. The initial concentration of the contaminant was from 25 to 100 mg/L and the catalyst loading was from 0.5 to 1 g/L for batch experiments. Depending on the initial concentration of dye, contaminant morphology, and catalyst content, the contaminant removal efficiency varied between 30 and 70%. However, according to the initial concentration of the pollutant and the irradiation time, a maximum removal efficiency of up to 98% was also reported in this photoreactor.

Kuo et al. [69] used a thin-film photoreactor with incorporated coating of TiO_2 nanoparticles with platinum and silver to remove o-cresol pollution as depicted in Figure 3.6. The results showed that the Pt-TiO_2 coatings and Ag-TiO_2 coatings improved the removal efficiency of the contaminant and that the efficiency was higher for Pt-TiO_2 coating compared with Ag-TiO_2 coating and pure TiO_2 nanoparticles.

Saran et al. [70] used a thin film plate reactor (TFPR) to purify the effluent from the secondary treatment of a sugar industry plant with sunlight on a pilot

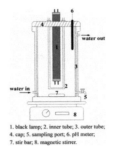

1. black lamp; 2. inner tube; 3. outer tube;
4. cap; 5. sampling port; 6. pH meter;
7. stir bar; 8. magnetic stirrer.

Figure 3.6 Schematic of thin film reactor. Reproduced from [69].

scale. In this study, Ag/TiO$_2$ nanoparticles were immobilized on ceramic tiles and the efficiency of treatment was compared to that obtained by using pure TiO$_2$ nanoparticles without silver. The results showed that the chemical oxygen demand (COD) removal efficiencies after 8 hours of sunlight by using Ag/TiO$_2$, pure TiO$_2$ nanoparticles and without the use of catalysts were 95%, 86%, and 22%, respectively. Also, the optimal values of sunlight duration, flow rate, initial pH, and the amount of H$_2$O$_2$ were 3 hours, 15 L/h, 2 and 5 mM, respectively, which resulted in COD removal as high as 99%. Coupling such a system with photovoltaic cells could provide the electricity needed for the circulation pump. The system, as shown in Figure 3.7, was able to purify the wastewater to such an extent that it could theoretically be used for some industrial purposes.

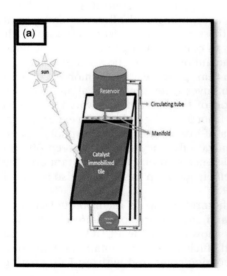

Figure 3.7 Pilot set-up of a thin film plate reactor reporting COD removal as high as 99%. Reproduced from [70].

3.3.4 Membrane reactors

Photocatalytic reactors using membranes are usually employed in three ways: (I) a slurry photoreactor chamber followed by a separate membrane module for filtration, (II) membranes used within a slurry photoreactor to remove the treated liquid without any escape of the catalyst, and (III) the use of immobilized catalysts on the membrane instead of employing a slurry. For the third option, instead of fabricating membranes from scratch, in some cases it is possible to purchase normal membranes which are available on the market and immobilize the photocatalytic catalyst on their surface.

Disadvantages of membrane photocatalytic reactors are: light scattering in the slurry catalyst (and any other disadvantage of the slurry reactor previously mentioned), difficult maintenance, high capital and operating costs, as well as the effect of wastewater components, such as particulate matter, mineral salts, and soluble organic pollutants, on membrane clogging and fouling [16]. Some variations of membrane photoreactors are shown in Figure 3.8.

3.3.5 Cascade reactors

These types of reactors consist of several similar stages in series so that in each stage, the output flow from the previous stage is treated. In other words, cascade reactors are not a specific reactor per se, but rather a combination of the other types of photocatalytic reactors previously explained.

In its simplest form, a cascade reactor can be a series of thin film plate reactors placed one after another where the outlet of one plate falls on the next plate and so on. As a case in point, Azadi *et al.* [71] studied the treatment of diluted landfill leachate with COD = 550 mg/L using a cascade photoreactor. The designed cascade reactor consisted of four steps made of Plexiglass with dimensions of 15 × 10 cm and a step height of 7.5 cm for each step. A glass

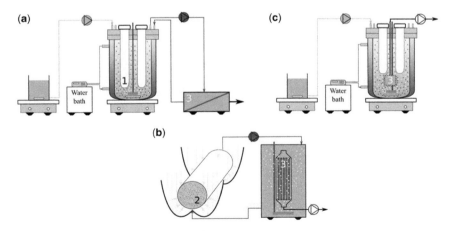

Figure 3.8 Scheme of a separated membrane photoreactor (a), a modified separated membrane photoreactor (b), an integrated membrane photoreactor (c); a lamp (1), a compound parabolic collector (2) and a membrane module (3). Reproduced from [16].

plate coated with carbon and tungsten titanium oxide (W-C-codoped TiO_2) nanoparticles was placed on each step. Two tanks with volumes of 1.8 and 0.125 L were placed at the bottom and top of the reactor, respectively, and the effluent was returned from the lower tank to the upper tank using a magnetic pump. The results showed that after 40 hours of treatment under 40 W intensity, coating surface density of 10.59 g/m², and leachate return rate of 1 L/min, COD removal efficiency of 84% was obtained.

Along the same lines, photocatalytic removal of benzoic acid was investigated by Chan et al. [72] using a cascade photoreactor consisting of three stainless steel plates as shown in Figure 3.9. TiO_2 nanoparticles were immobilized on the plates and UV lamps were used for irradiation. According to the results, the cascade reactor performed better than the single-plate reactor with a length and surface area comparable to the summation of all the used plates. It was postulated that in this type of configuration, the falling water improves the efficiency by increasing mixing and the transfer of oxygen into the solution. Such findings have been reaffirmed by Guillard et al. [73] for the decomposition of 4-chlorophenol and formetanate using a cascading photoreactor.

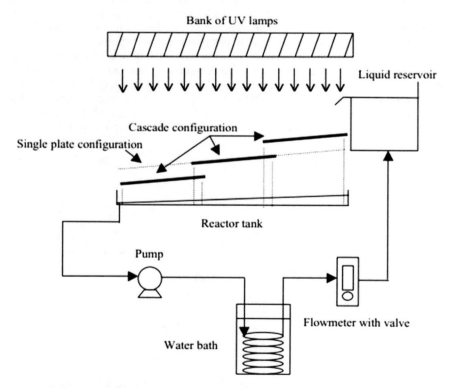

Figure 3.9 Schematic of a cascading reactor using several plates to improve mixing and oxygen transfer into the wastewater. Reproduced from [72].

Amiri et al. [74] used a cascade disk photocatalytic reactor coated with ZnO nanoparticles to investigate the effect of artificial roughness and flow rate on the photocatalytic degradation of reactive yellow 81 (RY81) pollutants. As evident from Figure 3.10, in order to set up this photoreactor, four circular disks with a radius of 34 cm were placed around a UV-C lamp as a light source. Sewage flow was pumped from the supply tank to the highest disk and from there it flowed downwards with a cascading flow regime through the holes in each disk (12 small holes of 4 mm diameter). These holes were embedded in a zigzag pattern to increase the time for the sewage stream to be exposed to light and to carry oxygen. The results showed that by creating artificial roughness on the disks and increasing the flow rate, the mass transfer coefficients increased which had a significantly positive effect on the removal of the aforementioned contaminant.

3.3.6 Spinning disk reactors

The differentiating component of a spinning disk reactor is that the catalyst is coated onto a disk which is subsequently mounted on a rotating motor with controllable speed. In this type of reactor, the wastewater flow is transferred by a pump to the center of the spinning disk and while flowing on it radially, it is irradiated by a light source. The output flow is collected at the bottom of

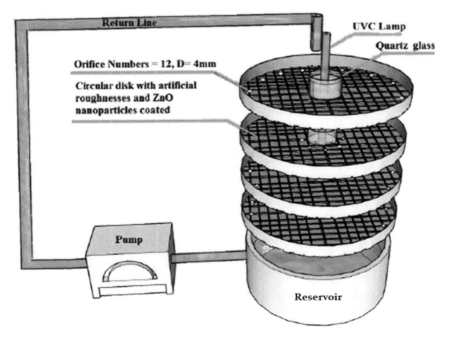

Figure 3.10 Schematic design of a photocatalytic cascading disk reactor. Reproduced from [74].

the reactor and can be returned to the beginning of the process if once-through treatment is not enough. After a period of time, the output flow concentration reaches steady state and the desired data can be prepared and analyzed. In previous studies, the components of the reactor have been covered with aluminum foil in order to prevent photolysis where it is unwanted [75, 76].

Yatmaz *et al.* [77] used a spinning disk reactor to degrade 4-chlorophenol and salicylic acid contaminants. The disk used in the reactor was made of borosilicate glass and TiO_2 nanoparticles were coated on it (Figure 3.11). The solutions of each contaminant flowed on the disk and were irradiated with a UV lamp. In this study, two types of low-pressure and medium-pressure mercury lamps were used (Figure 3.12), with peak wavelengths of UV rays emitted from the lamps being 254 and 365 nm, respectively. The results showed that the efficiency of the photocatalytic process depended on the type of UV source used and the shorter wavelength UV lamp had the highest removal efficiency.

In 2010, TiO_2 nanoparticles were immobilized onto a spinning disk which was then subjected to UV-C irradiation to remove methyl orange [78]. The diameter of the rotating disk used was 6 cm. The results showed that despite the short residence time of no more than a few seconds, a removal efficiency of more than 50% was achievable. In 2017, Mirzaei *et al.* [79] improved the performance of spinning disk reactors to remove phenols by using several baffles. They compared the removal of phenol contaminants using smooth and baffled spinning disk reactors as depicted in Figure 3.13. The results showed a significantly positive effect of the presence of baffles on the mixing and mass transfer, so that under

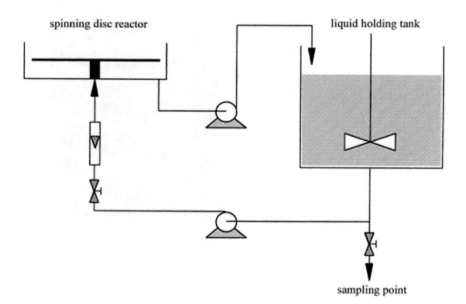

Figure 3.11 Layout of an experimental spinning disk set-up. Reproduced from [77].

Photocatalytic reactor types and configurations

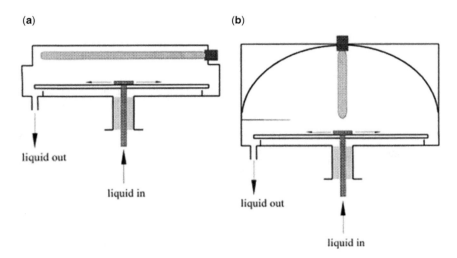

Figure 3.12 Lamp configurations used for the spinning disk reactor with arrows indicating direction of liquid flow; (a) 2 × 15 W low pressure mercury lamps, (b) 400 W medium pressure mercury lamp. Reproduced from [77].

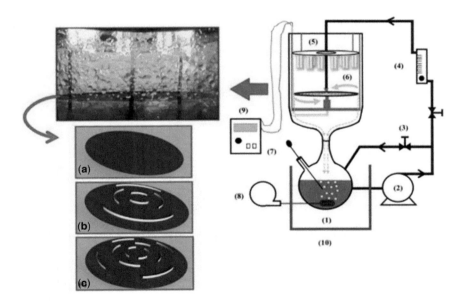

Figure 3.13 Schematic of spinning disk reactors with baffles which markedly improve performance. Reproduced from [79]. (1) feed reservoir, (2) pump, (3) valve, (4) flow meter, (5) strip of UV lamps, (6) spinning disc, (7) temperature sensor, (8) compressor, (9) power supplier, (10) circulating water bath. With different structures of disc surface: (a) smooth disc (b) baffled disc-I (c) baffled disc-II.

the same conditions, complete destruction of the contaminant in the baffled type occurred 90 minutes earlier than in the smooth type.

3.3.7 Rotating photocatalytic reactors

Rotating photocatalytic reactors have three main components: the reactor tube, the catalyst-coated disk, and the light source. This type of reactor consists of a series of coaxial disks coated with photocatalytic particles that are placed at a certain distance from each other in a tank. Any number of disks together form one stage, and a rotating photocatalytic reactor may consist of one or more stages. A portion of each of the disks is immersed in the reaction liquid and the rest of the disks, which are left outside of the reaction medium, are in contact with the oxygen in the atmosphere and are simultaneously exposed to UV or sunlight. The disks carry a thin film of liquid while spinning and emerging from the bath at a controllable speed. Under irradiation, the contaminants in the liquid film are destroyed. The concept behind rotating photocatalytic reactors is similar to that of rotating biological contactors which have been traditionally used in wastewater treatment [80, 81]. A schematic of the process is displayed in Figure 3.14.

Montalvo-Romero *et al.* [81] used a four-stage rotating photocatalytic reactor with ceramic disks coated with silver-modified TiO_2 nanoparticles for photocatalytic degradation of acetaminophen. The tank capacity was 50 L and the rotation speed of the disks was 54 rpm. The specifications of the reactor used are presented in Table 3.5. The kinetic results were consistent with the Langmuir model and showed a high degradation rate of the pollutant at low concentrations.

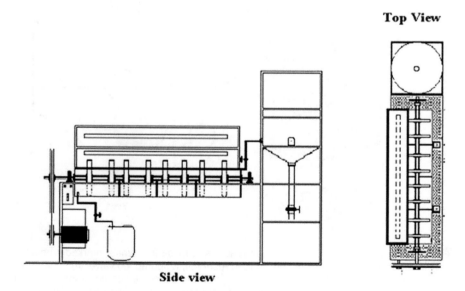

Figure 3.14 Schematic of a rotating photocatalytic reactor. Reproduced from [81].

Table 3.5 Characteristics of the system used in [81] for photocatalytic degradation of Acetaminophen.

Parameter	Specifications
Structure	Semi-cylindrical cover of stainless steel
Number of lamps	6 UV lamps (254 nm)
Disks	Assembled on one axis (stainless steel, diameter: 0.25 inches, length: 0.7 m)
	Material: ceramic (diameter: 0.24 m, thickness: 0.006 m, area per stage: 0.04523904 m^2)
Volume	3 L for each stage
Catalyst	TiO_2 doped with silver (5%)
Speed rotation	54 rpm

In another study, the photocatalytic degradation of 4-chlorobenzoic acid as an organic compound was studied using a rotating disk photoreactor with a rotation speed of 4 rpm, an initial solution volume of 3.5 L, under ambient temperature and near UV radiation [82]. The results further showed the possibility of destroying organic pollutants with such systems.

3.3.8 Taylor vortex photoreactor

The main components of a Taylor vortex reactor are a light source and two coaxial cylindrical vessels. The inner cylinder rotates at a relatively high speed (e.g., 4000 rpm) and produces narrow vortices around the central cylinder known as Taylor or Taylor-Couette vortices. The nature of these vortices and the resulting degree of mixing depend on the rotation speed of the inner cylinder, the fluid properties, and the dimensions of the reactor (Figure 3.15) [83, 84]. There is a relatively small gap between the inner and outer surfaces of the cylinders and, while performing aeration operations to supply the required oxygen in this part (if necessary), the reaction mixture flows through. If the light source is inside the inner cylinder, this cylinder must be transparent to allow light photons to pass through, and if the light source is outside the reactor, the outer cylinder must be transparent.

Subramanian and Kannan [85] studied the photocatalytic removal of phenol using a Taylor vortex photoreactor. The UV lamp was placed as a light source inside the inner cylinder (Figure 3.16) and the effect of operating parameters such as aeration, catalyst dose, internal cylinder rotation speed, gap size between the cylinders, initial pollutant concentration, and irradiation mode (periodic or continuous) were investigated. The results showed that continuous aeration of the reactor contents, in addition to providing the oxygen needed to trap electrons, is necessary to maintain photocatalytic reactions. Also, in reactors with small and medium gaps between cylinders, aeration caused mixing of the reaction contents, while in reactors with large gaps between cylinders, more mixing while rotating the inner cylinder with the help of the engine had a positive effect on the photocatalytic removal rate. The effect of increasing the dose of catalyst used on the removal efficiency depended on

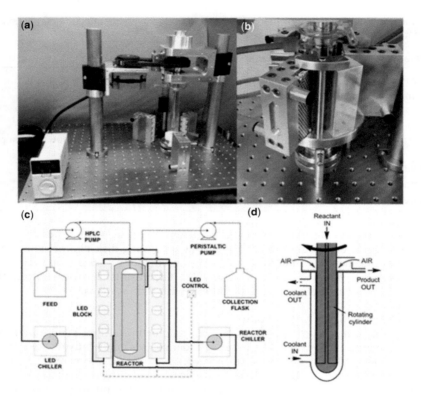

Figure 3.15 A Taylor vortex photoreactor; (a) showing the deconstructed reactor with the motor and its control box. A drive belt connects the motor and the rotating cylinder. During operation a protective housing (not shown) contains the motor, belt, and moving parts; (b) showing the LEDs and mirror blocks mounting positions around the reactor; (c) diagram of the reactor set-up showing the tubing connected to the reactor. Cooling is provided to the reactor by a recirculating chiller. The 3 LED blocks are connected in series and are cooled by a separate recirculating chiller; (d) A cross-section (not to scale) of the reactor showing the delivery and removal of reagents and air. Image reproduced from [84]. (HPLC: high performance liquid chromatography).

the mixing of the contents so that if sufficient mixing was provided, it had a positive effect and otherwise a negative effect. The periodic and regular rotation had no significant positive effect on reactor performance compared to the continuous mode.

3.3.9 Optical fiber reactors

Components of this type of reactor generally include a light source, optical fibers coated with photocatalytic particles, and a reaction vessel [1]. Optical fiber reactors have been used for photocatalytic degradation of various pollutants. In this regard, Danion *et al.* [86] investigated the photocatalytic degradation of

Photocatalytic reactor types and configurations

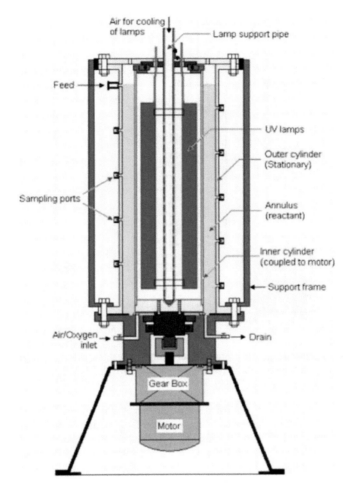

Figure 3.16 Schematic of an experimental set-up of a Taylor vortex photoreactor. Reproduced from [85].

malic acid with an initial concentration of 50 mg/L using a fiber optic reactor. To construct the reactor, a bundle consisting of 57 optical fibers (diameter 1 mm and length 30 cm) covered with five layers of TiO_2 with a thickness of 150 nm, inside a cylindrical container made of Pyrex glass with a volume of 220 mL (diameter 4.5 cm and length 13.5 cm) were used. In this reactor, a UV lamp was used as a light source to deliver the required light through the fiber bundles. The TOC removal efficiency was 21% after 20 h.

Lin et al. [87] used an optical fiber monolith reactor (Figure 3.17) to remove dichlorobenzene (DCB) and phenanthrene (PHE) from water. In this work, the

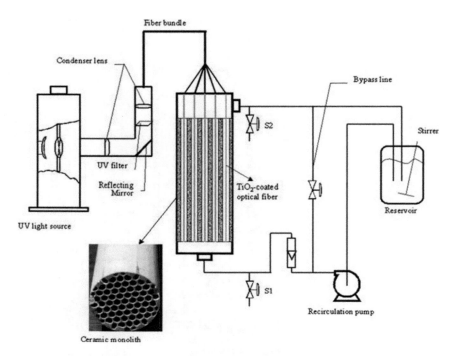

Figure 3.17 Schematic of the optical fiber monolith reactor experimental set-up. The reactor was in continuous recycle mode when it was tested. To measure the total removal efficiency, samples were obtained at S1 and S2. Reproduced from [87].

optimal TiO_2 layer thickness on the optical fiber was found to be 0.4 μm. DCB and PHE degradation kinetics were pseudo-first order. Also, results showed that overall removal efficiencies for DCB (initial concentration = 147 mg/L) and PHE (initial concentration = 505 μg/L) were 17.8% and 11.9%, respectively.

3.4 CONCLUDING REMARKS

The large number of studies on the design and use of photocatalytic reactors over the past years indicates the high potential of the photocatalytic method for use in water and wastewater treatment. In this chapter, the basic components in the design of photocatalytic reactors including light, catalysts, and reactors were investigated. Various forms of the configuration of photocatalytic reactors such as fluidized bed, thin film, cascading, annular, rotating disk, Taylor vortex, and optical fiber, and their relevant characteristics were examined.

Slurry systems provide a high specific surface area for photocatalytic particles. This leads to higher interaction between these particles and the reactants and consequently increases the treatment efficiency. However, limitations on light penetration and the need for post-treatment processes to separate photocatalytic particles from the treated stream, limit the possibility of commercialization and application of this type of reactor on a large scale. In

Table 3.6 Advantages and disadvantages of reactor types and configurations used in water and wastewater treatment.

Reactor Type	Advantages	Disadvantages	Catalyst	References
Parabolic trough reactor (PTR)	1. The intensity of the incoming light increases relative to the surface of the catalyst. 2. Photocatalyst consumption is reduced compared to photoreactors without light concentration. 3. Can be used at higher loading and pressure. 4. Low unit energy cost to increase the temperature.	1. Only absorbs and uses direct sunlight. 2. Problems on cloudy or dark days. 3. Increasing the intensity of the received light leads to limiting the reaction rate. 4. Factors such as the deformation of the collector due to wind and inaccuracy of the light detector cause errors.	Suspended catalyst	[88]
Composite parabolic collector (CPC)	1. More efficient light collection. 2. Can absorb and use direct and indirect sunlight. 3. No need for a sun tracker.	1. They need more photocatalysts than PTR. 2. The absorber tube is wider and therefore has lower operating pressure. 3. Unlike PTR, the surface area of the absorber tube in a decentralized CPC must be equal to the surface area of the reflector behind it.	Suspended catalyst	[89]
Oblique plate collectors	1. Relatively simple design. 2. Low cost. 3. No need to focus incoming light. 4. Able to absorb scattered light. 5. No need for a reflector. 6. It is possible to corrugate the surface of the reactor.	1. Difficult to create a thin layer of 100–200 µm. 2. Increasing the flow rate is difficult. 3. Direct exposure of its surface to the atmosphere leads to increased evaporation rates, interference by wind, and contamination with airborne particles, not to mention possible volatile organic compound (VOC) release.	Suspended and immobilized catalyst	[90]

(*Continued*)

Table 3.6 (*Continued*) Advantages and disadvantages of reactor types and configurations used in water and wastewater treatment.

Reactor Type	Advantages	Disadvantages	Catalyst	References
Rotating reactor	1. The mass transfer rate is high. 2. External oxygenation is not required. 3. Proper mixing is done. 4. The light intensity received is appropriate. 5. High efficiency for removing pollutants.	1. Relatively expensive set-up, requiring extra energy for rotation. 2. Excessive shear force may lead to catalyst escape.	Immobilized catalyst	[91]
Optical fiber reactors	1. Can collect distant lights. 2. Separation between light collection and photocatalytic materials.	1. Can only absorb and use direct light. 2. Problems on dark days. 3. Increasing the intensity of the received light leads to limiting the reaction rate. 4. Factors such as deformation of the collector due to wind and inaccuracy of the light detector can cause errors. 5. Low efficiency.	Immobilized catalyst	[80]
Fixed bed photoreactor	1. Catalyst escape is prevented to a large extent. 2. Oxygenation is achieved properly. 3. Good contaminant contact with the catalyst.	1. Light may not properly reach the catalyst. 2. The mass transfer rate is not very high. 3. Problematic implementation on a large industrial scale.	Immobilized catalyst	[92]

Table 3.7 A summary of studies of various reactor types for removing contaminants from water/wastewater.

Type of reactor	Contact area (m²)	Volume (L)	Photocatalyst Material	Amount	Type	Contaminant Chemical	Initial Concentration	Removal Efficiency (%)	Time/ Energy	Reference
Parabolic dishes	1	–	Pt-TiO₂	–	Immobilized	Toluene	20 ppm	79	5.7 s	[93]
	29.1	180	Ti-TiO₂	0.2 g/L	Slurry	4-Chlorophenol	27 mg/L	15	10 min	[80]
Combined parabolic	3.08	39	TiO₂	0.2 g/L	Slurry	Methyl orange	50 ppm	100	12.5 Einstein	[94]
Inclined plate	0.4	14	TiO₂	20 g/m²	Immobilized	Escherichia coli	1 000 000 CFU/mL	99	90 min	[95]
	0.06	0.5	ZnO	48 g/m²	Immobilized	Remazol red B	200 µM	83	99.5 min	[96]
	1.36	25	TiO₂	–	Immobilized	Formetanate	50 mg/L	100	20 kJ/L	[90]
	2	10	TiO₂	–	Immobilized	Landfill Leachate	COD = 26 000 mg/L	77	120.54 W.h	[18]
Rotating	0.014	0.025	Pt-TiO₂	11.4 g/m²	Immobilized	Phenol	16.5 mg/L	100	1.6 h	[97]
	–	–	TiO₂	40 g/m²	Immobilized	4-Chlorobenzoic acid	0.288 mmol/L	100	200 min	[98]
	–	–	TiO₂	–	Immobilized	Hydrocarbon	3500 ppm	92	300 min	[91]
	–	0.3	TiO₂	30 g	Slurry	Methylene blue	10 µM	35	90 min	[99]
	0.1	0.5	ZnO	–	Immobilized	Methyl orange	20 µM	90	100 min	[100]
Packed bed	–	–	TiO₂	–	Immobilized on activated carbon	Microcystin-LR (MC-LR)	50 mg/L	72	4 h	[92]
	–	0.13	TiO₂	5.48 g	Immobilized on glass bullets	Acid orange 7	30 mg/L	77	110 min	[58]
Optical fiber	0.025	4	TiO₂	–	Immobilized	4-Chlorophenol	0.1 mmol/L	100	13 h	[101]
	0.07	40	TiO₂	–	Immobilized	4-Chlorophenol	7.6 mg/L	79	8 h	[102]
Cascade	0.375	–	TiO₂	20 g/m²	Immobilized	Phenol	25 mg/L	96	300 min	[10]

contrast to slurry systems, immobilized photocatalytic reactors do not require post-processing or filtration. However, in immobilized reactors, the amount of contact between the contaminant and the catalyst surface and the mass transfer rate are less than in slurry systems, meaning that removal efficiencies can be lower. To overcome the limitation of mass transfer in these reactors, solutions such as stirring, rotation, cascading flow, or mixing via flow through reactor geometry can be used.

Regarding the different reactor types and configurations, fluidized bed reactors have very high efficiency, but if the goal is to use a natural light source such as sunlight, there are serious limitations in this regard. Inclined plate reactors are economically suitable for treating low flowrate effluents (<100 L/d) due to their simple design and the low cost of materials used in their construction. Widespread use of other configurations, such as rotating photocatalytic reactors, spinning disk reactors, and optical fiber reactors, are limited due to environmental sensitivity, mechanical complexity, high variability, difficulty in handling photocatalytic particles, and sometimes poor efficiency.

Table 3.6 provides a general comparison of the advantages and disadvantages of various reactors, including some which have not been thoroughly discussed in the chapter text; while Table 3.7 provides a list of studies pertaining to catalytic water and wastewater treatment with various reactors which might be of interest to readers. Although an attempt has been made to provide an overall view of various reactor types and configurations, it must be noted that photocatalytic reactors are in no way limited to those discussed in this chapter, and there is room for many innovative configurations in the years to come.

REFERENCES

1. Sundar KP, Kanmani S. Progression of Photocatalytic reactors and it's comparison: A Review. *Chemical Engineering Research and Design*, 2020; **154**: 135–150.
2. Lasa H De, Serrano B, Salaices M. *Photocatalytic Reaction Engineering*. Springer, Heidelberg, Germany, 2005.
3. Richards BS, Hudry D, Busko D, Turshatov A, Howard IA. Photon Upconversion for Photovoltaics and Photocatalysis: A Critical Review. *Chemical Reviews* 2021; **121**: 9165–95.
4. Marcelino RBP, Amorim CC. Towards visible-light photocatalysis for environmental applications: band-gap engineering versus photons absorption—a review. *Environmental Science and Pollution Research* 2019; **26**: 4155–70.
5. Wang W, Qi L. Light Management with Patterned Micro- and Nanostructure Arrays for Photocatalysis, Photovoltaics, and Optoelectronic and Optical Devices. *Advanced Functional Materials* 2019; **29**: 1807275.
6. Adams M, Skillen N, McCullagh C, Robertson PKJ. Development of a doped titania immobilised thin film multi tubular photoreactor. *Applied Catalysis B: Environmental* 2013; **130–131**: 99–105.
7. Kumordzi G, Malekshoar G, Yanful EK, Ray AK. Solar photocatalytic degradation of Zn2+ using graphene based TiO2. *Separation and Purification Technology* 2016; **168**: 294–301.
8. Zeghioud H, Khellaf N, Djelal H, Amrane A, Bouhelassa M. Photocatalytic Reactors Dedicated to the Degradation of Hazardous Organic Pollutants: Kinetics, Mechanistic Aspects, and Design – A Review. *Chemical Engineering Communications*, 2016; **203**(11): 1415–1431.

Photocatalytic reactor types and configurations 105

9. Al-Mamun MR, Kader S, Islam MS, Khan MZH. Photocatalytic activity improvement and application of UV-TiO2 photocatalysis in textile wastewater treatment: A review. *Journal of Environmental Chemical Engineering*, 2019; **7**(5): 103248.
10. Delnavaz M, Ayati B, Ganjidoust H, Sanjabi S. Kinetics study of photocatalytic process for treatment of phenolic wastewater by TiO2 nano powder immobilized on concrete surfaces. *Toxicological and Environmental Chemistry* 2012; **94**: 1086–98.
11. Li Puma G, Machuca-Martínez F, Mueses M, Colina-Márquez J. Scale-Up and Optimization for Slurry Photoreactors. In: Bustillo-Lecompte C, ed. *Advanced Oxidation Processes - Applications, Trends, and Prospects*. IntechOpen, London, UK, 2020.
12. Verma A, Tejo Prakash N, Toor AP, Bansal P, Sangal VK, Kumar A. Concentrating and Nonconcentrating Slurry and Fixed-Bed Solar Reactors for the Degradation of Herbicide Isoproturon. *Journal of Solar Energy Engineering, Transactions of the ASME* 2018; **140**(2): 021006.
13. Barzegar MH, Sabzehmeidani MM, Ghaedi M *et al.* S-scheme heterojunction g-C3N4/TiO2 with enhanced photocatalytic activity for degradation of a binary mixture of cationic dyes using solar parabolic trough reactor. *Chemical Engineering Research and Design* 2021; **174**: 307–18.
14. Rincón AG, Pulgarin C. Use of coaxial photocatalytic reactor (CAPHORE) in the TiO2 photo-assisted treatment of mixed E. coli and Bacillus sp. and bacterial community present in wastewater. *Catalysis Today*. Vol101. Elsevier, 2005: 331–44.
15. Vilhunen S, Sillanpää M. Recent developments in photochemical and chemical AOPs in water treatment: a mini-review. Springer, Heidelberg, Germany, doi 10.1007/s11157-010-9216-5.
16. Janssens R, Mandal MK, Dubey KK, Luis P. Slurry photocatalytic membrane reactor technology for removal of pharmaceutical compounds from wastewater: Towards cytostatic drug elimination. *Science of the Total Environment*, 2017; **599–600**: 612–626.
17. Mukhopadhyay S, Maiti D, Chatterjee S, Devi PS, Suresh Kumar G. Design and application of Au decorated ZnO/TiO2 as a stable photocatalyst for wide spectral coverage. *Physical Chemistry Chemical Physics* 2016; **18**: 31622–33.
18. Chemlal R, Abdi N, Drouiche N, Lounici H, Pauss A, Mameri N. Rehabilitation of Oued Smar landfill into a recreation park: Treatment of the contaminated waters. *Ecological Engineering* 2013; **51**: 244–8.
19. Chen X, Zhang F, Wang Q *et al.* The synthesis of ZnO/SnO2 porous nanofibers for dye adsorption and degradation. *Dalton Transactions* 2015; **44**: 3034–42.
20. Zhang Q, Fan W, Gao L. Anatase TiO2 nanoparticles immobilized on ZnO tetrapods as a highly efficient and easily recyclable photocatalyst. *Applied Catalysis B: Environmental* 2007; **76**: 168–73.
21. Rosy PJ, Jas MJS, Santhanalakshmi K, Murugan M, Manivannan P. Expert development of Hetero structured TiS2–TiO2 nanocomposites and evaluation of electron acceptors effect on the photo catalytic degradation of organic Pollutants under UV-light. *Journal of Materials Science: Materials in Electronics 2021 32:4* 2021; **32**: 4053–66.
22. Babic K, Tomašic V, Gilja V *et al.* Photocatalytic degradation of imidacloprid in the flat-plate photoreactor under UVA and simulated solar irradiance conditions—The influence of operating conditions, kinetics and degradation pathway. *Journal of Environmental Chemical Engineering* 2021; **9**: 105611.
23. Dantas SLA, Silva MMS, Gomes YF, Lopes-Moriyama AL, Souza CP, Corrêa MA. Photocatalytic degradation tests with cobalt-doped molybdenum carbides. *Applied Physics A 2021 127:2* 2021; **127**: 1–7.
24. Tahir MB, Ahmad A, Iqbal T, Ijaz M, Muhammad S, Siddeeg SM. Advances in photo-catalysis approach for the removal of toxic personal care product in

106 **Photocatalytic Water and Wastewater Treatment**

aqueous environment. *Environment, Development and Sustainability* 2020; **22**: 6029–52.

25. Rueda-Marquez JJ, Levchuk I, Fernández Ibañez P, Sillanpää M. A critical review on application of photocatalysis for toxicity reduction of real wastewaters. *Journal of Cleaner Production* 2020; **258**: 120694.

26. United States Patent 5116582 Photocatalytic slurry reactor having turbulence generating means. 1990 Apr 26.

27. Kim Y, Joo H, Her N, Yoon Y, Lee CH, Yoon J. Self-rotating photocatalytic system for aqueous Cr(VI) reduction on TiO2 nanotube/Ti mesh substrate. *Chemical Engineering Journal* 2013; **229**: 66–71.

28. de Lasa H, Serrano B, Salaices M. Novel Photocatalytic Reactors for Water and Air Treatment. In: de Lasa H, Serrano B, Salaices M, eds. *Photocatalytic Reaction Engineering*. Springer US, 2005: 17–47.

29. Sivagami K, Vikraman B, Krishna RR, Swaminathan T. Chlorpyrifos and Endosulfan degradation studies in an annular slurry photo reactor. *Ecotoxicology and Environmental Safety* 2016; **134**: 327–31.

30. Kamble SP, Sawant SB, Pangarkar VG. Photocatalytic degradation ofm-dinitrobenzene by illuminated TiO2 in a slurry photoreactor. *Journal of Chemical Technology & Biotechnology* 2006; **81**: 365–73.

31. Nishio J, Tokumura M, Znad HT, Kawase Y. Photocatalytic decolorization of azo-dye with zinc oxide powder in an external UV light irradiation slurry photoreactor. *Journal of Hazardous Materials* 2006; **138**: 106–15.

32. Wu CH, Chang HW, Chern JM. Basic dye decomposition kinetics in a photocatalytic slurry reactor. *Journal of Hazardous Materials* 2006; **137**: 336–43.

33. Tokumura M, Tawfeek Znad H, Kawase Y. Modeling of an external light irradiation slurry photoreactor: UV light or sunlight-photoassisted Fenton discoloration of azo-dye Orange II with natural mineral tourmaline powder. *Chemical Engineering Science* 2006; **61**: 6361–71.

34. Tasbihi M, Ngah CR, Aziz N *et al.* Lifetime and regeneration studies of various supported TiO2 photocatalysts for the degradation of phenol under UV-C light in a batch reactor. *Industrial and Engineering Chemistry Research*. Vol46. American Chemical Society, 2007: 9006–14.

35. Wang B, Karthikeyan R, Lu XY, Xuan J, Leung MKH. High photocatalytic activity of immobilized TiO2 nanorods on carbonized cotton fibers. *Journal of Hazardous Materials* 2013; **263**: 659–69.

36. Lei Z, Lee WI, Profile S *et al.* Mesoporous TiO2 photocatalytic films on stainless steel for water decontamination Mesoporous TiO2 photocatalytic films on stainless steel for water decontamination. *Cite this: Catal. Sci. Technol* 2012; **2**: 147–55.

37. Ahmed MH, Keyes TE, Byrne JA, Blackledge CW, Hamilton JW. Adsorption and photocatalytic degradation of human serum albumin on TiO2 and Ag-TiO2 films. *Journal of Photochemistry and Photobiology A: Chemistry* 2011; **222**: 123–31.

38. Li Y, Zhou X, Chen W *et al.* Photodecolorization of Rhodamine B on tungsten-doped TiO2/activated carbon under visible-light irradiation. *Journal of Hazardous Materials* 2012; **227-228**: 25–33.

39. Khatamian M, Hashemian S, Yavari A, Saket M. Preparation of metal ion (Fe 3+ and Ni 2+) doped TiO2 nanoparticles supported on ZSM-5 zeolite and investigation of its photocatalytic activity. *Materials Science and Engineering B: Solid-State Materials for Advanced Technology* 2012; **177**: 1623–7.

40. Sakthivel S, Shankar M V., Palanichamy M, Arabindoo B, Murugesan V. Photocatalytic decomposition of leather dye. Comparative study of TiO2 supported on alumina and glass beads. *Journal of Photochemistry and Photobiology A: Chemistry* 2002; **148**: 153–9.

Photocatalytic reactor types and configurations 107

41. Wang X, Shi F, Huang W, Fan C. Synthesis of high quality TiO2 membranes on alumina supports and their photocatalytic activity. *Thin Solid Films* 2012; **520**: 2488–92.
42. Dhananjeyan MR, Kiwi J, Thampi KR. Photocatalytic performance of TiO2 and Fe2O3 immobilized on derivatized polymer films for mineralisation of pollutants. *Chemical Communications* 2000: 1443–4.
43. Matthews RW. Photooxidative degradation of coloured organics in water using supported catalysts. TiO2 on sand. *Water Research* 1991; **25**: 1169–76.
44. Hosseini SN, Borghei SM, Vossoughi M, Taghavinia N. Immobilization of TiO2 on perlite granules for photocatalytic degradation of phenol. *Applied Catalysis B: Environmental* 2007; **74**: 53–62.
45. van Grieken R, Marugán J, Sordo C, Pablos C. Comparison of the photocatalytic disinfection of E. coli suspensions in slurry, wall and fixed-bed reactors. *Catalysis Today* 2009; **144**: 48–54.
46. Li D, Zheng H, Wang Q *et al.* A novel double-cylindrical-shell photoreactor immobilized with monolayer TiO2-coated silica gel beads for photocatalytic degradation of Rhodamine B and Methyl Orange in aqueous solution. *Separation and Purification Technology* 2014; **123**: 130–8.
47. Behnajady MA, Modirshahla N, Daneshvar N, Rabbani M. Photocatalytic degradation of an azo dye in a tubular continuous-flow photoreactor with immobilized TiO2 on glass plates. *Chemical Engineering Journal* 2007; **127**: 167–76.
48. Malakootian M, Nasiri A, Amiri Gharaghani M. Photocatalytic degradation of ciprofloxacin antibiotic by TiO2 nanoparticles immobilized on a glass plate. *Chemical Engineering Communications* 2020; **207**: 56–72.
49. Ye J, Li X, Hong J, Chen J, Fan Q. Photocatalytic degradation of phenol over ZnO nanosheets immobilized on montmorillonite. *Materials Science in Semiconductor Processing* 2015; **39**: 17–22.
50. Damodar RA, Swaminathan T. Performance evaluation of a continuous flow immobilized rotating tube photocatalytic reactor (IRTPR) immobilized with TiO2 catalyst for azo dye degradation. *Chemical Engineering Journal* 2008; **144**: 59–66.
51. Asha RC, Vishnuganth MA, Remya N, Selvaraju N, Kumar M. Livestock Wastewater Treatment in Batch and Continuous Photocatalytic Systems: Performance and Economic Analyses. *Water, Air, & Soil Pollution* 2015; **226**: 132.
52. Wang X, Jia J, Wang Y. Combination of photocatalysis with hydrodynamic cavitation for degradation of tetracycline. *Chemical Engineering Journal* 2017; **315**: 274–82.
53. Lee H, Park SH, Park Y-K *et al.* Photocatalytic reactions of 2,4-dichlorophenoxyacetic acid using a microwave-assisted photocatalysis system. *Chemical Engineering Journal* 2015; **278**: 259–64.
54. Divya N, Bansal A, Jana AK. Degradation of acidic Orange G dye using UV-H2O2 in batch photoreactor. *ajol.info* 2009.
55. Chanathaworn J, Bunyakan C, Wiyaratn W, Chungsiriporn J. *Photocatalytic Decolorization of Basic Dye by TiO2 Nanoparticle in Photoreactor.*
56. Colombo E, Ashokkumar M. Comparison of the photocatalytic efficiencies of continuous stirred tank reactor (CSTR) and batch systems using a dispersed micron sized photocatalyst. *RSC Advances* 2017; **7**: 48222–9.
57. Behnajady MA, Siliani-Behrouz E, Modirshahla N. Combination of design equation and kinetic modeling for a batch-recirculated photoreactor at photooxidative removal of C.I. Acid red 17. *International Journal of Chemical Reactor Engineering* 2012; **10**.
58. Sheidaei B, Behnajady MA. Mathematical kinetic modelling and representing design equation for a packed photoreactor with immobilised TiO2-P25 nanoparticles on

glass beads in the removal of C.I. Acid Orange 7. *Chemical and Process Engineering - Inzynieria Chemiczna i Procesowa* 2015; **36**: 125–33.

59. Tang C, Chen V. The photocatalytic degradation of reactive black 5 using TiO2/UV in an annular photoreactor. *Water Research* 2004; **38**: 2775–81.

60. Alam M, Mukhlish M, Uddin S, ... SD-J of S, 2012 undefined. Photocatalytic Degradation of Reactive Yellow in Batch and Continuous Photoreactor Using Titanium Dioxide. *banglajol.info*.

61. Adishkumar S, Kanmani S, Rajesh Banu J, Tae Yeom I. Evaluation of bench-scale solar photocatalytic reactors for degradation of phenolic wastewaters. *Desalination and Water Treatment* 2016; **57**: 16862–70.

62. Alexiadis A, Mazzarino I. Design guidelines for fixed-bed photocatalytic reactors. *Chemical Engineering and Processing: Process Intensification* 2005; **44**: 453–9.

63. Cloteaux A, Gérardin F, Thomas D, Midoux N, André JC. Fixed bed photocatalytic reactor for formaldehyde degradation: Experimental and modeling study. *Chemical Engineering Journal* 2014; **249**: 121–9.

64. Vezzoli M, Farrell T, Baker A, Psaltis S, Martens WN, Bell JM. Optimal catalyst thickness in titanium dioxide fixed film reactors: Mathematical modelling and experimental validation. *Chemical Engineering Journal* 2013; **234**: 57–65.

65. Vaiano V, Sacco O, Pisano D, Sannino D, Ciambelli P. From the design to the development of a continuous fixed bed photoreactor for photocatalytic degradation of organic pollutants in wastewater. *Chemical Engineering Science* 2015; **137**: 152–60.

66. Borges M, García D, Hernández T, Ruiz-Morales J, Esparza P. Supported Photocatalyst for Removal of Emerging Contaminants from Wastewater in a Continuous Packed-Bed Photoreactor Configuration. *Catalysts* 2015; **5**: 77–87.

67. Kabir MF, Vaisman E, Langford CH, Kantzas A. Effects of hydrogen peroxide in a fluidized bed photocatalytic reactor for wastewater purification. *Chemical Engineering Journal* 2006; **118**: 207–12.

68. Damodar RA, Jagannathan K, Swaminathan T. Decolourization of reactive dyes by thin film immobilized surface photoreactor using solar irradiation. *Solar Energy* 2007; **81**: 1–7.

69. Kuo YL, Su TL, Chuang KJ, Chen HW, Kung FC. Preparation of platinum-and silver-incorporated TiO2 coatings in thin-film photoreactor for the photocatalytic decomposition of o-cresol. *Environmental Technology* 2011; **32**: 1799–806.

70. Saran S, Kamalraj G, Arunkumar P, Devipriya SP. Pilot scale thin film plate reactors for the photocatalytic treatment of sugar refinery wastewater. *Environmental Science and Pollution Research* 2016; **23**: 17730–41.

71. Azadi S, Karimi-Jashni A, Javadpour S, Amiri H. Photocatalytic treatment of landfill leachate using cascade photoreactor with immobilized W-C-codoped TiO2 nanoparticles. *Journal of Water Process Engineering* 2020; **36**: 101307.

72. Chan AHC, Porter JF, Barford JP, Chan CK. Photocatalytic thin film cascade reactor for treatment of organic compounds in wastewater. *Water Science and Technology*. Vol**44**. IWA Publishing, 2001: 187–95.

73. Guillard C, Disdier J, Monnet C *et al.* Solar efficiency of a new deposited titania photocatalyst: Chlorophenol, pesticide and dye removal applications. *Applied Catalysis B: Environmental* 2003; **46**: 319–32.

74. Amiri H, Ayati B, Ganjidoust H. Mass transfer phenomenon in photocatalytic cascade disc reactor: Effects of artificial roughness and flow rate. *Chemical Engineering and Processing - Process Intensification* 2017; **116**: 48–59.

75. Boiarkina I, Norris S, Patterson DA. The case for the photocatalytic spinning disc reactor as a process intensification technology: Comparison to an annular reactor for the degradation of methylene blue. *Chemical Engineering Journal* 2013; **225**: 752–65.

Photocatalytic reactor types and configurations 109

76. Boiarkina I, Pedron S, Patterson DA. An experimental and modelling investigation of the effect of the flow regime on the photocatalytic degradation of methylene blue on a thin film coated ultraviolet irradiated spinning disc reactor. *Applied Catalysis B: Environmental* 2011; **110**: 14–24.
77. Yatmaz HC, Wallis C, Howarth CR. The spinning disc reactor - Studies on a novel TiO2 photocatalytic reactor. *Chemosphere* 2001; **42**: 397–403.
78. Chang CY, Wu NL. Process analysis on photocatalyzed dye decomposition for water treatment with TiO2-coated rotating disk reactor. *Industrial and Engineering Chemistry Research* 2010; **49**: 12173–9.
79. Mirzaei M, Jafarikojour M, Dabir B, Dadvar M. Evaluation and modeling of a spinning disc photoreactor for degradation of phenol: Impact of geometry modification. *Journal of Photochemistry and Photobiology A: Chemistry* 2017; **346**: 206–14.
80. Braham RJ, Harris AT. Review of major design and scale-up considerations for solar photocatalytic reactors. *Industrial and Engineering Chemistry Research*, 2009; **48**(19): 8890–8905.
81. Montalvo-Romero C, Aguilar-Ucán C, Alcocer-Dela hoz R, Ramirez-Elias M, Cordova-Quiroz V. A Semi-Pilot Photocatalytic Rotating Reactor (RFR) with Supported TiO2/Ag Catalysts for Water Treatment. *Molecules* 2018; **23**: 224.
82. Dionysiou DD, Balasubramanian G, Suidan MT, Khodadoust AP, Baudin I, Laîné JM. Rotating disk photocatalytic reactor: development, characterization, and evaluation for the destruction of organic pollutants in water. *Water Research* 2000; **34**: 2927–40.
83. Lee DS, Sharabi M, Jefferson-Loveday R, Pickering SJ, Poliakoff M, George MW. Scalable Continuous Vortex Reactor for Gram to Kilo Scale for UV and Visible Photochemistry. *Organic Process Research and Development* 2020; **24**: 201–6.
84. Lee DS, Amara Z, Clark CA *et al.* Continuous Photo-Oxidation in a Vortex Reactor: Efficient Operations Using Air Drawn from the Laboratory. *Organic Process Research and Development* 2017; **21**: 1042–50.
85. Subramanian M, Kannan A. Photocatalytic degradation of phenol in a rotating annular reactor. *Chemical Engineering Science* 2010; **65**: 2727–40.
86. Danion A, Disdier J, Guillard C, Abdelmalek F, Jaffrezic-Renault N. Characterization and study of a single-TiO2-coated optical fiber reactor. *Applied Catalysis B: Environmental* 2004; **52**: 213–23.
87. Lin H, Valsaraj KT. Development of an optical fiber monolith reactor for photocatalytic wastewater Treatment. *Journal of Applied Electrochemistry 2005 35:7* 2005; **35**: 699–708.
88. Fuqiang W, Ziming C, Jianyu T, Yuan Y, Yong S, Linhua L. Progress in concentrated solar power technology with parabolic trough collector system: A comprehensive review. *Renewable and Sustainable Energy Reviews*, 2017; **79**: 1314–1328.
89. Fernández P, Blanco J, Sichel C, Malato S. Water disinfection by solar photocatalysis using compound parabolic collectors. *Catalysis Today.* Vol101. Elsevier, 2005: 345–52.
90. Thu HB, Karkmaz M, Puzenat E, Guillard C, Herrmann JM. From the fundamentals of photocatalysis to its applications in environment protection and in solar purification of water in arid countries. *Research on Chemical Intermediates.* Vol31. Springer, 2005: 449–61.
91. Son H-J, Jung C-W, Kim S-H. Removal of Bisphenol-A using Rotating Photocatalytic Oxidation Drum Reactor (RPODR). *Environmental Engineering Research* 2008; **13**: 197–202.
92. Matsuda S, Hatano H. Photocatalytic removal of NOx in a circulating fluidized bed system. *Powder Technology.* Vol151. Elsevier, 2005: 61–7.
93. Sano T, Negishi N, Takeuchi K, Matsuzawa S. Degradation of toluene and acetaldehyde with Pt-loaded TiO2 catalyst and parabolic trough concentrator. *Solar Energy* 2004; **77**: 543–52.

94. Augugliaro V, Baiocchi C, Prevot AB *et al*. Azo-dyes photocatalytic degradation in aqueous suspension of TiO2 under solar irradiation. *Chemosphere* 2002; **49**: 1223–30.
95. Sichel C, Blanco J, Malato S, Fernández-Ibáñez P. Effects of experimental conditions on E. coli survival during solar photocatalytic water disinfection. *Journal of Photochemistry and Photobiology A: Chemistry* 2007; **189**: 239–46.
96. Selva Roselin L, Rajarajeswari GR, Selvin R, Sadasivam V, Sivasankar B, Rengaraj K. Sunlight/ZnO-Mediated photocatalytic degradation of reactive red 22 using thin film flat bed flow photoreactor. *Solar Energy* 2002; **73**: 281–5.
97. Zhang L, Kanki T, Sano N, Toyoda A. Photocatalytic degradation of organic compounds in Aqueous solution by a TIO2-Coated rotating-drum reactor using solar light. *Solar Energy* 2001; **70**: 331–7.
98. Dionysiou DD, Burbano AA, Suidan MT, Baudin I, Laîné JM. Effect of oxygen in a thin-film rotating disk photocatalytic reactor. *Environmental Science and Technology* 2002; **36**: 3834–43.
99. Salu OA, Adams M, Robertson PKJ, Wong LS, McCullagh C. Remediation of oily wastewater from an interceptor tank using a novel photocatalytic drum reactor. *Desalination and Water Treatment* 2011; **26**: 87–91.
100. Wang Q, Chen X, Yu K, Zhang Y, Cong Y. Synergistic photosensitized removal of Cr(VI) and Rhodamine B dye on amorphous TiO2 under visible light irradiation. *Journal of Hazardous Materials* 2013; **246–247**: 135–44.
101. Peill NJ, Hoffmann MR. Solar-powered photocatalytic fiber-optic cable reactor for waste stream remediation. *Journal of Solar Energy Engineering, Transactions of the ASME* 1997; **119**: 229–36.
102. Xu J, Ao Y, Fu D *et al*. Photocatalytic activity on TiO2-coated side-glowing optical fiber reactor under solar light. *Journal of Photochemistry and Photobiology A: Chemistry* 2008; **199**: 165–9.

doi: 10.2166/9781789061932_0111

Chapter 4
Landfill leachate treatment using photocatalytic methods

Hamed Sargazi[1], Alireza Bazargan[1]*, Meysam Shaghaghi[2] and Mika Sillanpää[3]

[1]School of Environment, College of Engineering, University of Tehran, Tehran, Iran
[2]School of Chemical Engineering, College of Engineering, University of Tehran, Tehran, Iran
[3]Department of Chemical Engineering, School of Mining, Metallurgy and Chemical Engineering, University of Johannesburg, P. O. Box 17011, Doornfontein 2028, South Africa
*Corresponding author: alireza.bazargan@ut.ac.ir

ABSTRACT

The growing global population and wasteful behavior of humans has led to ever-increasing production of solid wastes. The most common method for waste management on a global scale is landfilling. One of the main problems associated with landfills is the production of highly toxic liquid wastewater known as leachate, containing many organic and inorganic contaminants. The amount of leachate produced per ton of solid waste, as well as the exact characteristics of the leachate, are highly dependent on various issues such as weather, type of buried waste, and age of the landfill. If not properly collected, managed, and treated, leachate can have serious consequences for the environment. Various methods for leachate treatment have been proposed, one of which is the use of advanced oxidation processes (AOPs). Photocatalytic treatment, which is itself a subset of AOPs, is investigated in this chapter. For this purpose, after examining the quantitative and qualitative characteristics of leachate, homogeneous and heterogeneous photocatalytic processes for leachate treatment have been reviewed and relevant studies have been scrutinized. Various operational parameters such as pH, initial concentration of contaminants, intensity and absorption of light, dissolved oxygen, and photocatalyst dose, as well as different configurations of photocatalytic reactors have been shown to influence treatment effectiveness. In some cases, the use of photocatalysis for leachate treatment has shown considerable promise.

4.1 INTRODUCTION

With population growth, increased consumerism, and higher standards of living, the amount of municipal solid waste (MSW) generated across the globe is increasing. The total amount of MSW produced annually in the world has

© 2022 The Editors. This is an Open Access book chapter distributed under the terms of the Creative Commons Attribution Licence (CC BY-NC-ND 4.0), which permits copying and redistribution for noncommercial purposes with no derivatives, provided the original work is properly cited (https://creativecommons.org/licenses/by-nc-nd/4.0/). This does not affect the rights licensed or assigned from any third party in this book. The chapter is from the book Photocatalytic Water and Wastewater Treatment, Alireza Barzagan (Ed.).

112   **Photocatalytic Water and Wastewater Treatment**

been estimated to have been more than one billion tons in 2019, and according to Hoomweg and Bhada (2012), it will exceed two billion tons per year by 2025 [1]. In order to manage the large amount of MSW that is produced every day in urban areas, various management methods including incineration, composting, and landfilling are used, among which the use of landfills is the most common method. According to Gao *et al.* (2015), more than 95% of the municipal solid waste collected worldwide is disposed of in landfills [2] although in developed countries, there is a push to move towards more sustainable options such as resource recovery (circular economy).

One of the major problems associated with landfills is the production of a highly toxic liquid called leachate. Landfill leachate is defined as the wastewater produced by the infiltration of rainwater into the wastes buried in the landfill, as well as the flow of moisture in the waste, as well as liquid produced by biological and chemical reactions [3–6]. In one study it was estimated that for every ton of solid waste buried in a landfill, roughly 0.2 m³ of leachate is produced [7]. Of course, the amount of leachate is highly dependent on various factors such as waste characteristics and weather conditions, and this number is just an inaccurate approximation. To minimize the volume of leachate produced, various techniques, such as applying covers on landfills and waterproof layers at the sidewall and bottom of the landfill, are used [8].

Landfill leachate can pose a high risk to the environment due to its various pollutants, including organic and inorganic contaminants, heavy metals, and so on. In addition to soil and groundwater contamination, the horizontal movement of leachate from landfills and its exit from the soil surface at low altitudes causes surface water pollution. If such contaminated waters are used by the public utility for providing urban populations with drinking water, this can lead to the spread of disease and unwanted health risks [9, 10].

4.2 LANDFILL LEACHATE CHARACTERISTICS

Landfill leachate can be evaluated both quantitatively and qualitatively. The volume and quality of leachate produced directly depends on factors such as buried waste density, precipitation, evapotranspiration, surface runoff, infiltration, entry of groundwater into the landfill, chemical composition of the waste and age of the landfill [1, 10–12]. Over time, as the landfill ages, the leachate quality also changes [13].

According to previous studies, about 200 dangerous contaminants have been identified in landfill leachate so far [14]. These pollutants vary in quantity and quality in different landfills and have cumulative, threatening and detrimental effects on the food chain, resulting in many public health problems such as toxicity, carcinogenicity, mutagenicity, and teratogenicity [15–17]. This has made leachate an extremely complex wastewater for conventional treatment [18].

The various compounds in leachate can be divided into some general categories as follows [19]:

(1) Soluble organic matter (volatile fatty acids and recalcitrant organic compounds, such as humic acid and fulvic acid)

Landfill leachate treatment using photocatalytic methods 113

(2) Macro-mineral compounds including ammonia nitrogen ($N\text{-}NH_4^+$), sodium (Na^+), potassium (K^+), calcium (Ca^{2+}), magnesium (Mg^{2+}), manganese (Mn^{2+}), iron (Fe^{2+}), chloride (Cl^-), sulfate (SO_4^{2-}) and hydrogen carbonate (HCO_3^-)

(3) Heavy metals such as chromium (Cr^{3+}), nickel (Ni^{2+}), copper (Cu^{2+}), zinc (Zn^{2+}), cadmium (Cd^{2+}), mercury (Hg^{2+}) and lead (Pb^{2+}), as well as rare earth elements

(4) Xenobiotic organic compounds such as cyclic hydrocarbons, phenols, chlorinated aliphatics, pesticides and lubricants, as well as oils.

The quality of landfill leachates is generally determined by a series of physicochemical parameters, including: pH, suspended solids (SS), 5-day biological oxygen demand (BOD_5), chemical oxygen demand (COD), NH_4^+, total nitrogen (TN), chloride, phosphorus, heavy metals and alkalinity [11, 18, 20]. As mentioned, one of the determining factors of the quality of leachate, and consequently the amount of the above parameters, is the age of the landfill. Depending on the age, common landfill leachate can be divided into three or more categories [19]:

(1) Young (<5 years)
(2) Intermediate (5–10 years)
(3) Old leachate (>10 years)

As landfill ages, several leachate parameters such as pH, BOD_5, COD, and BOD_5/COD ratio change significantly [21, 22]. The composition of the leachate varies not only depending on the age of the landfill, but also from place to place, and this causes large fluctuations in the values of these parameters. Table 4.1, after considering hundreds of articles and studies, expresses the concentrations of different compounds in leachate at various ages. Table 4.2 presents the values of some quality parameters in some landfill leachate samples worldwide based on landfill age. Meanwhile, Figure 4.1 shows the typical values of pH, COD and BOD_5/COD ratio for young, intermediate and mature landfill leachate [23, 24].

The concentration of contaminants is high at the beginning of leachate formation and decreases continuously over time (after years) along with the stabilization of the waste material. As the age of the landfill increases, BOD_5 and COD concentrations decrease, due to the decomposition of organic matter in the leachate [27]. Most biodegradable organic matter is thought to degrade during the stabilization phase, but the non-biodegradable organic matter remains [28]. As a result, the BOD_5/COD (biodegradability index) ratio decreases over time. High concentrations of COD (> 10 000 mg/L) and BOD_5/COD ratio between 0.5 to 1 are observed in young landfill leachate, while COD concentration less than 4000 mg/L and BOD_5/COD ratios less than 0.1 are found in mature landfill leachate [21, 22]. On the other hand, ammonia is another major contaminant in landfill leachate. In general, with increasing age of the landfill, the concentration of ammonia nitrogen does not show a clear decreasing trend [18, 21].

According to Figure 4.2, leachate formation is the result of several biological processes, including hydrolysis, aerobic degradation (phase 1), and anaerobic degradation (phase 2–4).

Table 4.1 Characteristics of leachate during landfill life [25].

Parameters	Young	Intermediate	Stabilized	Old
Age (years)	5>	5–10	10–20	>20
pH	3–7	8	>7.5	>7.5
Alkalinity	1000–20 000	500–6000	–	–
Conductivity (μS/cm)	2000–50 000	1000–15 000	–	–
Organic compounds	80% VFA	5–30% VFA + fulvic acid and humic acid	Fulvic acid and humic acid	Fulvic acid and humic acid
BOD_5 (mg/L)	2000–50 000	500–15 000	50–1000	<300
COD (mg/L)	4000–90 000	1000–30 000	1000–5000	<3000
BOD_5/COD	0.5–1	0.1–0.5	<0.1	<0.1
TOC/COD	<0.3	0.3–0.5	>0.5	>0.5
Heavy metals (mg/L)	>2	–	–	<2
SO_4^{2-} (mg/L)	300–4000	100–2000	20–200	<100
P (mg/L)	50–500	5–200	–	<20
Cl (mg/L)	500–6000	200–4000	50–500	<200
Zn^{2+} (mg/L)	50–400	20–200	5–50	<20
Fe^{3+} (mg/L)	200–3000	200–2000	50–500	<200
Ca^{2+} (mg/L)	1000–7000	200–4000	100–500	<400
Na^+ (mg/L)	1000–7000	200–3000	50–500	<200
Mg^{2+} (mg/L)	300–3000	200–2000	50–500	<200
TKN (mg/L)	500–4500	400–2000	50–2000	<2000
NH_3-N (mg/L)	<400	–	>400	–
NH_4^+-N (mg/L)	500–4500	–	–	<1500

VFA: volatile fatty acid; TOC: total organic carbon; TKN: total Kjeldahl nitrogen.

At the solid liquid interface, as long as oxygen is available, aerobic bacteria can perform hydrolysis rather quickly. Hydrolysis leads to the breakdown of polysaccharides into monosaccharides. Furthermore, lipids are broken down into fatty acids and glycerol, and proteins break down into their building blocks, i.e. proteins [29]. The concentration of oxygen continues to diminish until it reaches zero within a few days. This means that the production of CO_2, H_2O, nitrate and sulfate through aerobic decomposition does not continue for long.

Anaerobic degradation can itself be divided into three stages: anaerobic acidification, unstable methanogenesis, and stable methanogenesis (stabilization). The first stage involves acid fermentation, which results in the production of CO_2 and volatile fatty acids and reduces the pH to 5.5 or less. The second stage begins with the slow growth of methanogenic bacteria. The high levels of volatile fatty acids produced in the first stage inhibit the initial metabolism of these bacteria. The conversion of CO_2 and volatile acids to methane, the total duration of which varies from a few months to 1–2 years, leads to a gradual increase in pH

Landfill leachate treatment using photocatalytic methods 115

Table 4.2 The leachate composition of some important landfills around the world [26].

Country	NH_4^+-N	TN [g/m³]	pH	BOD/COD	BOD [g/m³]	COD [g/m³]
Young Landfills						
Canada	42	212	5.8	0.7	9660	13800
China	2260	–	7.7	0.27	4200	15700
	1154	1384	8.0	0.66	5669	8528
Greece	3100	3400	6.2	0.38	26800	70900
Italy	3917	–	8	0.2	4000	19900
Turkey	2500	–	7.8	0.55	11000	20000
*Average	2162±1385	1665±1612	7.2±1	0.46±0.21	10221±8619	24805±22982
Intermediate Landfills						
China	–	–	7.6	0.07	430	5800
Germany	840	1135	–	0.33	1060	3180
Greece	940	1100	7.9	0.2	1050	5350
Hong Kong	1500	2000	6.4	0.24	1600	6610
Italy	1130	–	8.4	0.25	1270	5050
Poland	743	–	8	0.28	331	1180
Turkey	1270	1450	8.1	–	–	9500
Average	1070±285	1421±416	7.7±0.7	0.23±0.09	957±490	5239±2618
Old Landfills						
China	1972	2117	8.5	–	–	1703
	1040	1055	8.8	–	–	1819
	2000	2030	8	0.05	100	2200
Finland	159	192	–	0.11	62	556
France	430	540	7.7	0.01	7.1	500
Germany	445	–	–	0.26	290	1225
Greece	238	357	8.4	–	–	2456
Poland	340	420	8.1	0.07	51	732
Spain	1623	2199	–	0.12	558	4512
	3772	4058	8.1	0.19	810	4357
	5975	6317	8.5	0.03	123	3921
	2003	2305	8.3	0.13	832	6200
Vietnam	3449	3868	8.4	0.12	425	3621
Average	1616±1557	1939±1715	8.2±0.3	0.12±0.07	332±304	2652±1786

*The mean values calculated in this study are based on the values reported in previous studies±standard deviation.

[26]. In the third stage (stable methanogenesis), methane production continues until the complete consumption of biodegradable organic matter is achieved. This portion of the process may take as long as 15–20 years. Following the stable methanogenic stage, with the complete degradation of biodegradable organic matter, the amount of generated CH_4 and CO_2 gases diminishes [29].

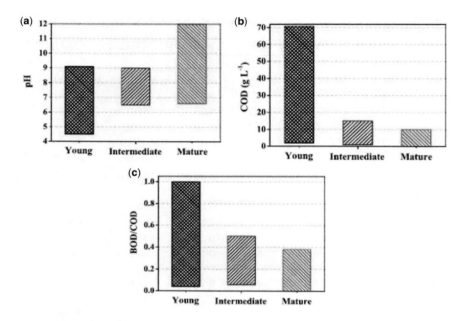

Figure 4.1 Typical values of leachate parameters (a) pH (b) COD and (c) BOD5/COD ratio in young, intermediate, and mature landfills Reprinted from [21].

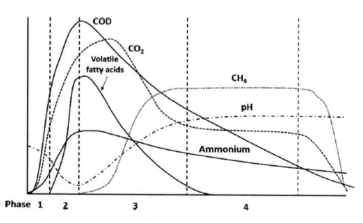

Figure 4.2 Landfill phases: 1. Aerobic phase; 2. Anaerobic acidification phase; 3. Unstable methanogenic phase; 4. Stable methanogenic phase Reprinted from [26].

4.3 LANDFILL LEACHATE TREATMENT

Conventional landfill leachate treatment solutions can be categorized into four main groups [10, 27, 30, 31]:

(1) Biological treatment processes such as aerobic or anaerobic treatment.
(2) Physical and chemical processes including chemical oxidation, adsorption, filtration, chemical precipitation, coagulation and flocculation, flotation and precipitation.
(3) Thermal processes, including both natural and forced evaporation and incineration [32–35].
(4) Indirect solutions, such as recirculating the leachate back into the sanitary landfill, or transferring it to a wastewater treatment plant to be treated alongside municipal wastewater.

Usually, a combination of solutions are employed and a leachate treatment plant has numerous treatment stages including biological and physicochemical processes. The performance of some conventional landfill treatment processes based on landfill age are presented in Table 4.3. Of course, these recommendations are very general, and the efficiency of processes will vary for each specific landfill and leachate.

According to the above table, since the BOD_5/COD ratio is high (0.5–1) in young landfill leachate, biological methods are particularly effective in the early years. However, some landfill leachates may be toxic to microorganisms, even if they have seemingly suitable BOD_5/COD ratios. It is also worth noting that biological processes are suitable for biodegradable organic pollutants, while in leachate, resistant organic pollutants are also present. In old landfill leachate, the BOD_5/COD ratio is low (about 0.1), which indicates that most of the organic

Table 4.3 Performance of different treatment processes based on landfill age [19].

	Age of Landfill (Years)		
Leachate Treatment Method	**Young (<5)**	**Intermediate (5–10)**	**Mature (>10)**
Co-treatment with domestic wastewater	Good	Fair	Poor
Recycling	Good	Fair	Poor
Aerobic process (suspended growth)	Good	Fair	Poor
Aerobic process (fixed film)	Good	Fair	Poor
Anaerobic process (suspended growth)	Good	Fair	Poor
Anaerobic process (fixed film)	Good	Fair	Poor
Natural evaporation	Good	Good	Good
Coagulation/flocculation	Poor	Fair	Fair
Chemical precipitation	Poor	Fair	Poor
Adsorption	Poor	Fair	Good
Oxidation	Poor	Fair	Fair
Air stripping	Poor	Fair	Fair

pollutants in the leachate are resistant and non-biodegradable, and therefore biological processes alone are not efficient for treating mature leachate. In other words, it is not possible to successfully use biological processes during the entire lifetime of the landfill and a combination of approaches is required as the system ages [4].

This is where the possibility of using photocatalytic treatment comes in. One of the methods that has attracted the attention of researchers regarding the decomposition of resistant organic pollutants, which may prove very effective in leachate treatment in the future, is photocatalytic oxidation [4]. Among the advantages of this method are the following:

- Oxidation of organic matter to the highest degree of oxidation, that is, carbon dioxide and water
- The process can be performed at ambient temperature and pressure
- Improved biodegradability of recalcitrant organic pollutants such that they can be degraded in downstream biological processes

Conventional photocatalytic processes for leachate treatment are divided into two main categories: homogeneous systems with radiation ($H_2O_2/Fe^{2+}/$ UV (photo-Fenton)) and heterogeneous systems with radiation ($TiO_2/O_2/UV$, $TiO_2/H_2O_2/UV$) [36].

4.4 HOMOGENEOUS PHOTOCATALYTIC PROCESSES FOR LEACHATE TREATMENT

A homogeneous photocatalytic method for removing resistant contaminants in leachate is the photo-Fenton method [36]. The photo-Fenton method is based on the production of hydroxyl radicals from the decomposition of hydrogen peroxide in the presence of iron (II) ions [37]. Hydroxyl radicals react non-selectively with the constituents in the leachate to form more biodegradable intermediates [38–40].

Primo *et al.* (2008) investigated the leachate treatment in the Cantabria region of northern Spain. The average initial values of COD and BOD_5 of the leachate were 3823 and 680 mg/L, respectively. Initially, the leachate was biologically treated at the landfill site to reduce organic matter and ammonia, but this method alone could not meet the standard discharge limits of Spain (e.g., COD<160 mg/L). For this reason, the researchers investigated five advanced oxidation methods, photo-Fenton, Fenton-like, Fenton, UV/H_2O_2 and UV, to further treat the leachate and compared the efficiencies of these methods in removing organic charge and leachate color. By using the photo-Fenton method under optimal conditions with $H_2O_2 = 10\ 000$ mg/L and 2000 mg Fe^{2+}/L, 86% of the COD and more than 95% of the color were removed from the leachate [41]. The results showed that the photo-Fenton process had the highest removal efficiency, and the removal efficiency of the methods was ranked from high to low as: photo-Fenton>Fenton-like>Fenton>UV/H_2O_2>UV.

Although studies have reported relatively high degradation of organic matter using the photo-Fenton process [41–43], the results of a study by Hermosila *et al.* (2009) on COD removal efficiency using the Fenton and the photo-Fenton

Landfill leachate treatment using photocatalytic methods 119

processes showed a similar removal efficiency of 70% for both of these processes. Although the removal efficiencies were comparable, the main advantage of the photo-Fenton process over the Fenton process was the reduction of Fe^{2+} consumption (32 times less) and the volume of sludge produced (25 times less). Therefore, the additional cost associated with UV radiation in the photo-Fenton process can be offset by reducing the amount of Fe^{2+} ions required and reducing the amount of sludge produced, which makes the photo-Fenton process competitive with the Fenton process [44].

Poblete and Perez (2020) used sawdust as a pretreatment step in the photo-Fenton process along with ozonation to purify landfill leachate. The initial values of COD, ammonium, iron, nitrate, and copper in the leachate samples were 11,950, 4628.5, 55.8, 34.5, and 375 mg/L, respectively, with pH = 8.9. The results of pretreatment removal efficiency for ammonium, iron, copper, humic acid, COD, and color were 87%, 70.2%, 61.1%, 18.3%, 33.7%, and 19.5%, respectively. The photo-Fenton and ozonation were then applied and the results of total removal efficiency for COD, color, ammonium, and humic acid were increased to 95.1%, 95%, 94.5%, and 97.9%, respectively [45].

Li *et al.* (2019) investigated and compared photo-Fenton and photo-Fenton-like ($UV/Cu^{2+}/H_2O_2$) processes to treat mature landfill leachate. They used the response surface methodology (RSM) to optimize the reaction conditions. The photo-Fenton-like system performed well in the decomposition of organic matter and broke down resistant organic pollutants such as fulvic acid and humic acid, thereby improving the biodegradability of the landfill leachate. The maximum dissolved organic carbon (DOC) removal efficiency in the photo-Fenton process was higher than in the photo-Fenton-like process (74.21% vs. 65.23%). It is noteworthy that after treatment, the recovery capacity of the Cu^{2+} catalyst in the photo-Fenton-like process was higher than that of the Fe^{2+} catalyst in the photo-Fenton process [46]. The sensitivity of photo-Fenton and photo-Fenton-like processes to the pH parameter was reported to be different, and the $UV/Cu^{2+}/H_2O_2$ photo-Fenton-like process had optimal performance over a wider range of pH; whereas in contrast, the photo-Fenton process was more sensitive and showed optimal performance in a more limited range of pH [46].

Numerous studies have been performed to evaluate and compare the performance of Fenton and photo-Fenton processes for landfill leachate treatment. Regarding the COD parameter, the results showed that the COD removal efficiency was almost equal for both processes [47]. However, in some studies, the efficiency of COD removal in the photo-Fenton process was slightly higher than in the Fenton process. For example, Ghanbarzadeh Lak *et al.* (2018) reported a difference of 10% and reduced the COD of Tehran's leachate from 17 200 to 2970 mg/L by the photo-Fenton process under optimal conditions [48].

4.4.1 Combination of photo-Fenton methods with other treatment options

If it is decided to use biological methods for leachate treatment, in order to make the recalcitrant organic matter more biologically available for microorganisms, it is better to use a form of pretreatment. The photo-Fenton method as an AOP, has been successfully used for pretreatment of leachate before biological

120   Photocatalytic Water and Wastewater Treatment

treatment [49, 50]. Colombo *et al.* (2019) initially used separate photo-Fenton and biological methods to treat landfill leachate. After using the conventional biological process, the removal efficiencies of COD and BOD_5 were 87% and 84%, respectively. In the photo-Fenton process under the optimal conditions (retention time $= 120$ min, pH $= 4$, $[Fe^{2+}] = 80$ mg/L, and $[H_2O_2] = 3400$ mg/L) COD and BOD_5 removal efficiencies were 89% and 75%, respectively. The results showed that the rate of removal of COD and BOD_5 were low, and the resulting effluent was not clean enough to meet the standard discharge limits. To solve this problem, the researchers used a combined photo-Fenton/biological method. The combined method led to total removal efficiency of 98% for COD and BOD_5. It is noteworthy that the initial concentration was 12 797 mg/L for COD and 4251 mg/L for BOD_5 [51].

A pretreatment process that has been widely used *prior* to the photo-Fenton process is coagulation and flocculation. In this regard, several studies have been conducted, for example, Silva *et al.* (2016) showed that if the coagulation and flocculation process is used as a pretreatment prior to the photo-Fenton process, the amounts of consumed H_2O_2 and contact time in comparison to when no pretreatment is used are reduced by 7% and 17%, respectively [52]. Tejera *et al.* (2019) used a three-step combined method with the sequence of coagulation-flocculation, photo-Fenton, and biological sequencing batch reactor (SBR) method to treat landfill leachate. They performed this combined method in two ways and compared the results; first case: coagulation with ferric chloride + homogeneous photo-Fenton + SBR; second case: coagulation with alum coagulant + heterogeneous photo-Fenton with zero valent iron particles + SBR. In both cases, COD and color removal efficiency were reported above 90% and the final treated effluent met the standard allowable range. It is worth mentioning that in this study, the initial concentration of COD in the leachate was about 5000 mg/L. However, the economic evaluation showed that the second case is four times more expensive than the first case [53].

Silva *et al.* (2016) evaluated the photo-Fenton process from an economic point of view to treat biologically pretreated leachate on an industrial scale. The photo-Fenton process was carried out in three modes; A) using parabolic concentrators B) using UV lamps, and C) a combination of modes A and B, to reduce the amount of COD of the treated stream from the initial value of 2945–4864 mg/L to the two desired values of 1000 mg/L (for discharge to the sewage collection network) and 150 mg/L (for discharge to receiving water). However, it was concluded that in order to reach the value of COD<150 mg/L, after the photo-Fenton process, a biological treatment step must be used. The total unit costs to reach the COD concentration target $= 1000$ mg/L for set-ups A, B and C were: 6.8, 7.2, and 6.7 €/m³, respectively. For reaching the more stringent goal of COD $= 150$ mg/L the costs respectively increased to: € 11/m³, € 11.7/m³ and € 10.9/m³ [39].

Elsewhere, Li *et al.* (2016) studied the concentrated effluent from the nanofiltration stage of landfill leachate treatment in two existing treatment plants A and B. The initial COD concentrations in these two concentrated leachate samples were 1330 and 3450 mg/L, respectively. The combined Fenton-coagulation process, followed by the photo-Fenton process were used

Landfill leachate treatment using photocatalytic methods 121

Table 4.4 The optimal operational parameters and removal efficiency of leachate organic matter in some studies.

Type of Leachate	Initial COD (mg/L)	pH	H_2O_2 (mg/L)	Fe^{2+} (mg/L)	Contact Time (min)	Removal Efficiency of COD (%)	References
Raw	2527–2770	2.8	5066	60	–	–	[55]
Raw	12 797	2.4	3400	80	120	89	[56]
Stabilized	1670	–	3300	600	120	70.7	[47]
Biological pretreated	1685	2.4	3400	80	120	87	[51]
Raw	4123	–	3060	1680	90	68	[57]
Biological pretreated	3300–4400	3–3.5	–	2000	60	86	[41]
Raw	5200	2.8	2000	10	60	57.5	[58]

to treat the two streams. For sample A, the results showed a COD removal efficiency of about 70% using the Fenton-coagulation method and then 71% of the residual COD was removed using the photo-Fenton process. For sample B, these results were 68% and 85%, respectively. Also, the amount of BOD_5/COD in sample A increased from 0.03 in the as-received leachate to 0.17 following the Fenton-coagulation process and 0.35 after the photo-Fenton process. For sample B, this value increased from 0.01 to 0.04 and then to 0.43, respectively. This indicates the positive effects of the Fenton-coagulation and photo-Fenton processes to increase the biodegradability of concentrated leachate [54].

4.4.2 Operational parameters
In order to optimize the photo-Fenton process, various parameters are often tweaked. The considered parameters usually consist of: pH, reagent dose (Fe^{2+} and H_2O_2 concentration), injection method of agent, temperature, initial COD concentration of leachate, reaction time, and recycling of the produced sludge.

The optimal values of the mentioned parameters along with the organic matter removal efficiency of the leachate treated by the homogeneous photocatalytic process are presented in Table 4.4 for various studies.

4.5 HETEROGENEOUS PHOTOCATALYTIC PROCESS FOR LEACHATE TREATMENT
Another type of photocatalytic process used to remove persistent contaminants in landfill leachate is the heterogeneous photocatalytic process. One example is the heterogeneous photocatalytic degradation of COD and DOC of leachate by Jia *et al.* [59]. The experiment was carried out in a Pyrex cylindrical reactor with a volume of 1000 mL and TiO_2 nanoparticles were used. In order to evaluate the efficiency of treatment, the test period was extended to 72 hours and sampling was performed at 0, 6, 12, 24, 36, 48, 60, and 72 hours for analysis. The initial COD and DOC of the leachate in this study were 2440 and 914 mg/L,

respectively. At pH of 4–5 and the optimal concentration of TiO_2 (2 g/L), the removal efficiencies of COD and DOC were 60% and 74%, respectively [59].

In 2013, Chemlal *et al.* investigated the treatment of surface and groundwater contaminated with landfill leachate with a COD of about 1200– 34 000 mg/L using a photocatalytic process. The photocatalytic particles used were TiO_2 particles stabilized on sloping plates. The leachate stream was irradiated for 30 to 54 hours using three 15 W UV-C lamps while passing over the sloping plates. In this study, with a change in pH, the rate of COD removal was reported to be between 76 and 92% indicating that the initial pH of the effluent strongly influences the outcome [6].

In 2002, researchers conducted a study on leachate treatment using TiO_2 nanoparticles (Degussa P-25) as photocatalysts under UV irradiation to investigate the removal of COD, BOD_5, and TOC [60]. The leachate mainly contained non-biodegradable materials with a low BOD_5/COD ratio and the average initial COD was 1409 mg/L. A batch reactor was used and TiO_2 powder with anatase phase with an average diameter of 20 nm was used suspended in the leachate. For irradiation, a UV-C lamp with a wavelength of 254 nm and a power of 8 W was used. Optimal contaminant removal efficiencies were achieved in this experiment in 12 hours. By performing experiments at different pHs, the maximum removal efficiency was obtained at a pH of 4. By performing the experiments in the optimal conditions, the removal efficiencies of COD, BOD_5, and TOC were reported to be 59%, 75%, and 80%, respectively. The results also showed that removal of BOD and TOC is favored (is both faster and happens to a greater extent) under acidic conditions compared to an alkaline environment [60].

In another study, the treatment of landfill leachate by TiO_2 nanoparticles under UV radiation was again considered [61]. The purpose of this research was to investigate the effective factors in the photocatalytic process including the pH of the solution, the number of nanoparticles used, and the reaction time to determine the maximum removal of organic matter and color. During this process, the COD and color removal efficiencies were 60% and 97%, respectively, with the COD value decreasing from 2440 to 976 mg/L. Also, the ratio of BOD_5 to COD increased from 0.09 to 0.39, which indicates an increase in the biodegradability of the leachate [61].

One of the contaminants in landfill leachate is 1,4-dioxane, which is widely used as a solvent in the pharmaceutical, dyeing, and other industries. To remove this contaminant from biologically pretreated leachate, Nemura *et al.* (2020) investigated and compared a photocatalytic rotating advanced oxidation contactor (RAOC) coupled with AC/TiO_2 composite sheets with a TiO_2 slurry reactor. The results showed that because of strong inhibition by coexisting substances in the landfill leachate, the TiO_2 slurry reactor method had lower removal efficiency, but in the RAOC method, the 1,4-Dioxane removal efficiency was higher. This was due to the formation of a very thin film on the composite sheets and the adsorption of UV-absorbing materials on the film. It should be noted that in this study, the duration of the process in both methods was 66 hours [62].

4.5.1 Modification of nanocatalysts

One strategy for improving the performance of photocatalysts for leachate treatment has been to dope nano-photocatalysts, and thereby add a small number of impurities. Such defects in the structure of the nanoparticle can cause agitation of the catalyst in the presence of visible light, followed by improved optical efficiency. For example, Teh and Mohamed (2011) used holmium to contaminate TiO_2 to increase its photocatalytic effectiveness. This action produced more hydroxyl radicals due to the reduction of excited electron-holes recombination [63].

Elleuch *et al.* (2020) investigated the performance of a combined biological pretreatment method by kefir grains and a photocatalytic process using Ag-coated TiO_2 nanoparticles to remove toxic contaminants from landfill leachate [64]. The initial value of the COD was 24 000 mg/L. The results showed that if kefir grains with a concentration of 1% w/v were used for pretreatment operations at 37°C, the removal efficiencies of TOC, COD, NH_4^+-N, and PO_4^{3-} were 93%, 83.3%, 70% and 88.3%, respectively. Also, Cd, Ni, Zn, Mn, and Cu removal efficiencies were 100%, 94%, 62.5%, 53.2%, and 47.5%, respectively. In order to optimize the Ag-coated TiO_2 photocatalytic process, statistical analysis in the form of RSM with Box-Behnken design was used. With optimization the removal efficiencies of TOC and COD were increased to 98% and 96%, respectively, while the removal efficiency for all other parameters also increased. The results showed that TiO_2 doped with Ag along with kefir grains has a good potential for regeneration and reuse, which is an important parameter in the application of catalysts. For this reason, this combined biological-photocatalytic method was deemed as a suitable method for the treatment of leachate [64].

Shahmoradi *et al.* (2018) studied the treatment of leachate produced in composting operations with an initial COD concentration of 2000 mg/L by photocatalytic treatment using sunlight [65]. They used Nd-coated ZnO photocatalytic nanoparticles to reduce the band gap energy and transfer it to the visible range. Design Expert software with RSM was used for optimization purposes in order to improve the decomposition efficiency. Under optimal conditions of pH = 6.74, Nd-doped ZnO concentration of 1920 mg/L, reaction time of 114.62 minutes, and H_2O_2 concentration of 12.56 mmol/L, the COD removal efficiency was 82.19% [65].

Also, treatment of diluted landfill leachate with a COD concentration of 1000 mg/L by the heterogeneous photocatalytic method using TiO_2 nanoparticles under UV irradiation was studied by Thuong *et al.* (2015). COD and dye removal efficiencies were 67.2% and 56.4%, respectively, under optimal conditions of pH = 4, TiO_2 nanoparticle concentration of 0.05 g/L and UV irradiation time of 90 minutes [66].

More recently, Azadi *et al.* (2020) used a cascade photoreactor with carbon and tungsten-coated titanium oxide nanoparticles (W-C-codoped TiO_2) fixed on its inner plates to treat diluted landfill leachate at an initial concentration of COD = 550 mg/L. The results showed that after 40 hours of treatment under 40 W intensity, coating surface density of 10.59 g/m², and leachate return flow rate of 1 L/min, a COD removal efficiency of 84% was obtained [67].

4.5.2 Operational, structural and environmental parameters

The operational parameters frequently considered for the optimization of heterogeneous photocatalytic processes are pH, initial concentration of contaminant in the leachate, intensity and absorption of light, dissolved oxygen, reactor configuration, and the amount of photocatalyst used. In the following, a summary of the effects of these parameters is provided according to the literature.

4.5.2.1 pH

One of the most important operational parameters in the heterogeneous photocatalytic treatment of landfill leachate is pH, which can affect the surface charge and isoelectric point of catalytic particles, the size of the catalytic aggregate, the position of the valence band and conduction band, and the mechanism of production of hydroxyl radicals. In fact, due to differences in the distribution of molecular charges, the ease of bond breaking and the site of attack in photocatalysis may differ as the pH varies. pH influences the surface charge of semiconductors, interfacial electron transport, and photoredox reactions that occur in their presence [4]. Numerous studies have reported the effect of pH on the efficiency of the photocatalytic process. The study by Kashitarash *et al.* (2012) showed that increasing the pH to 6.5, increased the removal efficiency of COD, BOD_5, total solids (TS), and color. While a further increase of pH from 6.5 to 8.5 led to a decrease in the removal efficiency. The reason for this result was the occupation of active sites on the surface of nanoparticles by ferrous hydroxide deposition during the contact of iron ions with hydroxyl radicals [68]. According to Huang *et al.* (2020), the removal efficiency of chloride from landfill leachate decreased with increasing pH and the highest removal efficiency (59.3%) occurred at pH = 1 [69]. Also, landfill leachate treatment in a fixed bed photocatalytic reactor with a thin film using UV radiation for 30–54 hours was investigated. The maximum COD removal efficiency was obtained at an optimal pH of about 5 [6].

4.5.2.2 The initial concentration of contaminants in the leachate

With all other variables held constant, as the initial concentration of COD in the leachate increases, the COD removal becomes less effective. This is because with lower concentration of contaminants, more pollutants are exposed to photocatalytic particles. Other reasons as to why an increase in contaminant concentrations can lead to reduced removal efficiencies are as follows [67]:

- Reduction of the ratio of the number of active photocatalysts to the total number of pollutant molecules
- Reduction of the number of photons that can penetrate the solution and reach the surface of the catalysts, resulting in less production of reactive species
- Occupancy of active sites on the surface of the photocatalyst particles due to surface adsorption of pollutants on the surface of the particles and consequent reduction of the number of active reactants produced.

4.5.2.3 *The intensity and absorption of light*

The photocatalytic process begins and continues with the absorption of light and the subsequent production of electron-hole pairs on the surface of the photocatalyst particles. The rate of photocatalytic reaction increases to a certain extent with increasing light intensity. However, if the light intensity increases beyond the threshold, the photocatalytic reaction rate will not change much [4].

The effect of light intensity on the COD removal efficiency of leachate from compost was studied by Ranjbari and Mokhtarani (2018) using 8, 16, 24, 32, and 40 watts of UV radiation. The results showed that by increasing UV radiation from 8 to 16 W, the COD removal efficiency increased by about 15% while increasing the UV radiation from 32 to 40 W led to a 2% increase in COD removal efficiency [70]. Azadi *et al.* (2020) used a photocatalytic process to investigate the COD removal efficiency of leachate under three irradiations of 10, 20, and 40 watts for 60 hours, which resulted in COD removal efficiencies of 53%, 64%, and 67%, respectively. Since changing the light from 20 to 40 W only resulted in a minor increase in COD removal, 20 W was chosen as the optimal amount in order to reduce excessive energy usage [67]. Also, Mokhtarani *et al.* (2015) investigated UV light with values of 8, 16, 32, 47, 62, 77, 92 and 107 watts to remove COD of biologically pretreated leachate from a composting unit in northern Iran. The initial concentration of the COD was 450 mg/L and it was shown that the removal rate increased as the light power was increased to 77 W. After that, increasing the light power above 77 W maintained a consistent removal efficiency [71].

4.5.2.4 *Dissolved oxygen*

In photocatalytic reactions, dissolved oxygen (DO) plays an important role in trapping the excited electron of the conduction band. In fact, DO reacts with electrons, thereby reducing the rate of electron-hole recombination reactions [4, 71]. Jia *et al.* (2010) used an aeration system with a flow rate of 1.5 L/min to provide DO and maintain flow mixing in a reactor using UV/TiO_2 to increase contact between photocatalytic particles and pollutants. The experimental results showed that COD, DOC, and color removal by UV-TiO2 photocatalysis could be as high as 60%, 70%, and 97%, respectively [61].

4.5.2.5 *Reactor configuration*

Various configurations of slurry and stabilized photocatalytic reactors have been used to treat landfill leachate such as the annular, falling film flow reactor, slurry membrane, thin-film cascade, continuous flow, rotating drum, plate, baffle, and optical fiber photoreactors [72]. The details regarding some photoreactor configurations used for landfill leachate treatment are presented in Table 4.5.

4.5.2.6 *Photocatalyst concentration*

In slurry photoreactors, with the increase of photocatalytic particles, the density of particles per unit of irradiated surface increases, and as a result, the degradation efficiency of pollutants improves. Another way to frame this

Photocatalytic Water and Wastewater Treatment

Table 4.5 Types of reactors and their configuration in leachate treatment.

Reactor Type	Dimensions	Material	Volume (Liters)	References
Annular	–	Quartz	0.375	[73]
Shell and tube	–	Quartz/Quartz and steel	–	[74]
Tubular	–	Borosilicate glass	0.6	[75]
Cylindrical	90 mm inside diameter and 250 mm height	Stainless steel	1.25	[76]
Slurry	4.5 × 2.5 × 2.5 inches	Galvanized aluminum sheets	1	[77]
Cylindrical	12.5 cm inner diameter and 31.8 cm height	With aluminum coating	2.3	[78]

improvement is that an increase in the number of photocatalysts, increases the number of active catalytic sites to adsorb the reacting molecules and to produce more reactive free radicals. However, excessive amounts of photocatalysts should not be used. In addition to being cost prohibitive, over-concentration of photocatalysts can cause agglomeration of the photocatalytic particles which reduces the overall surface, and also hinderance of the penetration of light rays into the solution. Therefore, determination of the optimal amount of photocatalytic particles is required to achieve maximum efficiency of photocatalytic activity [79]. In this regard, Jia *et al.* (2011) studied the effect of TiO_2 nanoparticle concentration on photocatalytic removal of organic matter [61]. The results showed that, as expected, the amount of TiO_2 had a great effect on the removal efficiency of COD and color. It was found that the removal efficiency of the pollutants increased with increasing the concentration of TiO_2 to the value of 2 g/L, and then decreased as the concentration of the photocatalyst was increased beyond this point [61].

This observation is also true for photoreactors with stabilized photocatalytic particles. As the amount of coverage of photocatalytic nanoparticles on the reactor plates increases, the number of active sites on the surface increases and as a result, the COD removal efficiency in the leachate increases. However, as the coating thickness increases beyond a particular value, the surface porosity of the nanoparticles in the substrate decreases, the catalysts compress, and the internal mass transfer rate decreases. In the presence of nanoparticles, two simultaneous adsorption and photocatalytic processes occur. Increasing the thickness of the photocatalyst layer inhibits the adsorption mechanism. Therefore an optimal thickness exists for the photocatalyst film [80, 81]. For example, Mokhtarani *et al.* (2015) investigated the effect of stabilized TiO_2 on the removal efficiency of organic matter in the leachate, and performed experiments in the range of 5–90 g of photocatalyst per unit area. The operating parameters of the tests were pH = 5, reaction duration of 24 hours, using an 8 W UV-C lamp. The results showed that with increasing the amount of stabilized

Landfill leachate treatment using photocatalytic methods 127

catalyst, the removal efficiency increases and reaches a maximum value, and a further increase of the photocatalyst from the optimal level (60 g/m²) has almost no effect on the removal efficiency [71].

4.5.3 Combination of the heterogeneous photocatalytic method with other treatment methods

In some studies, the application of the photocatalytic method has been investigated as a pretreatment method in the leachate treatment processes. For example, Yasmin *et al.* (2020) investigated the application of the photocatalytic process using TiO_2/Ag nanocomposites as a pretreatment process for biological treatment. The removal efficiencies of COD, TOC, and NH_4^+-N in the pretreatment process under optimal conditions were 70%, 71.25%, and 49.1%, respectively. The efficiencies of the combined process (pretreatment + biological treatment) for COD, TOC, and NH_4^+-N were 90%, 85%, and 75%, respectively. In other words, the COD level decreased from 24 000 to 2400 mg/L. Besides, heavy metals such as Fe, Zn, Cu, Cd, and Pb were removed with efficiencies ranging from 50% to 95%. The results showed that TiO_2/Ag has high photocatalytic activity, and that the nanoparticles could be reused. After using the nanocomposites five times, the removal efficiency of organic pollutants was still above 60% [82].

In some studies, the photocatalytic process has been used as a secondary treatment following other treatments. In this regard, Mokhtar *et al.* (2010) used coagulation as a pretreatment followed by solar photocatalysis with ZnO nanoparticles. The coagulation process was performed under two pH values of 8.68 and 5 by adding 10 g $FeCl_3/L$ as a coagulant for the first case and 1 g/L in the second case. Color removal efficiencies were 67% and 35%, respectively. Then, by using 1 g ZnO nanoparticles per liter at a pH of 5 and a reaction duration of 120 minutes, a color removal efficiency of 97% was obtained. When a lower amount of ZnO was used (0.2 g/L) with the addition of 50 mL/L of H_2O_2 at pH = 5, the color removal efficiency was 95% [83].

Elsewhere, Hassan *et al.* (2016) used the UV/TiO_2 photocatalytic method to treat biologically pretreated mature landfill leachate. Under the operating conditions of pH = 5 and nanoparticle concentration of 1 g/L, a COD removal efficiency of 82% was obtained [84].

4.6 CONCLUSION

In recent years, among the various processes for leachate treatment, including physical, chemical, biological and a combination of treatment methods, the photocatalytic treatment method has been identified by researchers as a promising treatment option. Photocatalytic treatment can be carried out at ambient temperature and pressure, rendering recalcitrant organic pollutants more biodegradable, and possibly oxidizing pollutants and organic matter to the highest degree of oxidation, that is to carbon dioxide and water.

In young landfill leachate, the BOD_5/COD ratio is high, so biological methods for the removal of biodegradable organic contaminants have higher efficiency, while in more mature landfill leachate, recalcitrant organic contaminants are

Table 4.6 A summary of several studies using photocatalytic methods for leachate treatment.

Photocatalytic Reactants and Catalysts	Pollutant Type and Concentration in Feed Leachate (mg/L)	Pollutant Type and Concentration after Photocatalytic Treatment (mg/L)	Light Properties and Intensity (W)	Comments	References
TiO_2	COD (500 mg/L)	COD (340 mg/L)	Solar radiation	Optimal pH was equal to 6	[85]
P25 nano-TiO_2	COD (2440.3 mg/L), Color (2400 times)	COD (976.12 mg/L), Color (72 times)	15 W low-pressure UV mercury lamp $\lambda = 254$ nm	Contact time = 72 h and the initial pH and TiO_2 dosage are major factors controlling photocatalysis	[61]
TiO_2 (Degussa P–25)	COD (18 600 mg/L)	COD (7800 mg/L)	150 W UV-A, $\lambda = 365$ nm; Light intensity: 5 kW/m²	Contact time = 1 h	[75]
TiO_2 (Degussa P–25)	COD (706 mg/L), TOC (203 mg/L)	COD (104 mg/L), TOC (36.83 mg/L)	8 W UV-C mercury lamp $\lambda = 254$ nm	The role of DO was important in photocatalysis	[74]
W-doped TiO_2	COD (325 mg/L)	COD (175.5 mg/L)	2 × 36 W fluorescent lamp $\lambda = 400–700$ nm	optimal conditions: pH = 6.63 tungsten content = 2.64%	[86]
$WTiO_2$ (mixture of TiO_2 and Fe III)	COD (13 646 mg/L)	COD (1910 mg/L)	Solar UV irradiation $\lambda > 350$ nm; Light intensity: 30 W/cm²	The use of $WTiO_2$ produced a higher organic matter removal rate (86%) than photo-Fenton using commercial FeSO4 (43%)	[87]
TiO_2 (Degussa P–25)	COD (1409 mg/L)	COD (986 mg/L)	8 W UV-C, $\lambda = 254$ nm; Light intensity: 21 W/cm²	Under the acidic condition, the photocatalytic removal efficiency of landfill leachate was relatively high	[60]
W-doped TiO_2	COD (678 mg/L)	COD (316 mg/L)	2 × 36 W fluorescent lamp	Optimal conditions: pH = 6.3 tungsten content = 2.2%	[88]
W/C-codoped TiO_2	COD (2050 mg/L)	COD (615 mg/L)	UV-visible light	Optimal condition (W/C-codoped): pH = 7.17 W-C = 2.66–0.06% Calc. temp = 571°C	[89]
Ag-doped TiO_2 nanoparticles	COD (24 000 mg/L)	COD (960 mg/L)	UV light $\lambda < 385$ nm	They used a combination of biological pretreatment with kefir grains and photocatalytic process with Ag-doped TiO_2 nanoparticles	[64]

Landfill leachate treatment using photocatalytic methods

more prevalent. The ratio of BOD_5/COD is low (about 0.1) in mature landfill leachate, which indicates that most of the organic pollutants in this leachate are recalcitrant and non-biodegradable, and therefore biological processes are not suitable for the treatment of mature leachate. Hence, to degrade these pollutants into more biodegradable pollutants, the photocatalytic process can be used in combination with other treatment processes. A summary of the studies pertaining to the treatment of landfill leachate with photocatalytic methods is provided in Table 4.6.

Disadvantages of photocatalytic treatment methods include energy use, the required retention times, the sensitivity to operating parameters such as turbidity and so on. The catalytic particles may also adsorb pollutants while creating reactive oxidizing radicals. This is problematic because it hinders the reuse of the catalysts. Therefore, nanoparticle regeneration methods could be the subject of future research. Also, in homogeneous systems, it is necessary to use hydrogen peroxide during the treatment process. The transfer and storage of hydrogen peroxide adds a level of complexity to the application of the process. Therefore, using innovative methods to produce this substance at the point of consumption should be studied.

REFERENCES

1. Costa AM, Alfaia RG de SM, Campos JC. Landfill leachate treatment in Brazil – An overview. *Journal of Environmental Management*, 2019; **232**: 110–116.
2. Gao J, Oloibiri V, Chys M *et al.* The present status of landfill leachate treatment and its development trend from a technological point of view. *Reviews in Environmental Science and Bio/Technology* 2015; **14**: 93–122.
3. Salem Z, Hamouri K, Djemaa R, Allia K. Evaluation of landfill leachate pollution and treatment. *Desalination* 2008; **220**: 108–14.
4. Hassan M, Zhao Y, Xie B. Employing TiO2 photocatalysis to deal with landfill leachate: Current status and development. *Chemical Engineering Journal*, 2016; **285**: 264–275.
5. Cai FF, Yang ZH, Huang J, Zeng GM, Wang L ke, Yang J. Application of cetyltrimethylammonium bromide bentonite-titanium dioxide photocatalysis technology for pretreatment of aging leachate. *Journal of Hazardous Materials* 2014; **275**: 63–71.
6. Chemlal R, Abdi N, Drouiche N, Lounici H, Pauss A, Mameri N. Rehabilitation of Oued Smar landfill into a recreation park: Treatment of the contaminated waters. *Ecological Engineering* 2013; **51**: 244–8.
7. Kurniawan TA, Lo W hung. Removal of refractory compounds from stabilized landfill leachate using an integrated H2O2 oxidation and granular activated carbon (GAC) adsorption treatment. *Water Research* 2009; **43**: 4079–91.
8. Dajić A, Mihajlović M, Jovanović M, Karanac M, Stevanović D, Jovanović J. Landfill design: Need for improvement of water and soil protection requirements in EU Landfill Directive. *Clean Technologies and Environmental Policy* 2016; **18**: 753–64.
9. Reinhart D, Berge N, Management EB-S and HW, Report undefined, 2007 undefined. Long term treatment and disposal of landfill leachate.
10. Renou S, Givaudan JG, Poulain S, Dirassouyan F, Moulin P. Landfill leachate treatment: Review and opportunity. *Journal of Hazardous Materials*, 2008; **150**(3): 468–493.
11. Miao L, Yang G, Tao T, Peng Y. Recent advances in nitrogen removal from landfill leachate using biological treatments – A review. *Journal of Environmental Management*, 2019; **235**: 178–185.

12. El-Fadel M, Bou-Zeid E, Chahine W, Alayli B. Temporal variation of leachate quality from pre-sorted and baled municipal solid waste with high organic and moisture content. *Waste Management* 2002; **22**: 269–82.
13. Kulikowska D, Klimiuk E. The effect of landfill age on municipal leachate composition. *Bioresource Technology* 2008; **99**: 5981–5.
14. Clarke BO, Anumol T, Barlaz M, Snyder SA. Investigating landfill leachate as a source of trace organic pollutants. *Chemosphere* 2015; **127**: 269–75.
15. Toufexi E, Tsarpali V, Efthimiou I, Vidali MS, Vlastos D, Dailianis S. Environmental and human risk assessment of landfill leachate: An integrated approach with the use of cytotoxic and genotoxic stress indices in mussel and human cells. *Journal of Hazardous Materials* 2013; **260**: 593–601.
16. Gajski G, Oreščanin V, Garaj-Vrhovac V. Chemical composition and genotoxicity assessment of sanitary landfill leachate from Rovinj, Croatia. *Ecotoxicology and Environmental Safety* 2012; **78**: 253–9.
17. Budi S, Suliasih BA, Othman MS, Heng LY, Surif S. Toxicity identification evaluation of landfill leachate using fish, prawn and seed plant. *Waste Management* 2016; **55**: 231–7.
18. Kjeldsen P, Barlaz MA, Rooker AP, Baun A, Ledin A, Christensen TH. Present and long-term composition of MSW landfill leachate: A review. *Critical Reviews in Environmental Science and Technology*, 2002; **32**(4): 297–336.
19. Luo H, Zeng Y, Cheng Y, He D, Pan X. Recent advances in municipal landfill leachate: A review focusing on its characteristics, treatment, and toxicity assessment. *Science of the Total Environment*, 2020; **703**: 135468.
20. Bhatt AH, Karanjekar R V., Altouqi S, Sattler ML, Hossain MDS, Chen VP. Estimating landfill leachate BOD and COD based on rainfall, ambient temperature, and waste composition: Exploration of a MARS statistical approach. *Environmental Technology and Innovation* 2017; **8**: 1–16.
21. Fernandes A, Pacheco MJ, Ciríaco L, Lopes A. Review on the electrochemical processes for the treatment of sanitary landfill leachates: Present and future. *Applied Catalysis B: Environmental*, 2015; **176–177**: 183–200.
22. Foo KY, Hameed BH. An overview of landfill leachate treatment via activated carbon adsorption process. *Journal of Hazardous Materials*, 2009; **171**(1–3): 54–60.
23. Lo HM, Liao YL. The metal-leaching and acid-neutralizing capacity of MSW incinerator ash co-disposed with MSW in landfill sites. *Journal of Hazardous Materials* 2007; **142**: 512–9.
24. Wang ZP, Zhang Z, Lin YJ, Deng NS, Tao T, Zhuo K. Landfill leachate treatment by a coagulation-photooxidation process. *Journal of Hazardous Materials* 2002; **95**: 153–9.
25. Reshadi MAM, Bazargan A, McKay G. A review of the application of adsorbents for landfill leachate treatment: Focus on magnetic adsorption. *Science of the Total Environment*, 2020; **731**: 138863.
26. Bove D, Merello S, Frumento D, Arni S Al, Aliakbarian B, Converti A. A Critical Review of Biological Processes and Technologies for Landfill Leachate Treatment. Chem Eng Technol. Available. *Chemical Engineering and Technology* 2015; **38**(12): 2115–2126.
27. Torretta V, Ferronato N, Katsoyiannis I, Tolkou A, Airoldi M. Novel and Conventional Technologies for Landfill Leachates Treatment: A Review. *Sustainability* 2016; **9**: 9.
28. Ahmed FN, Lan CQ. Treatment of landfill leachate using membrane bioreactors: A review. *Desalination* 2012; **287**: 41–54.

Landfill leachate treatment using photocatalytic methods 131

29. Cossu R, Raga R. Test methods for assessing the biological stability of biodegradable waste. *Waste Management* 2008; **28**: 381–8.
30. Iskander SM, Zhao R, Pathak A *et al.* A review of landfill leachate induced ultraviolet quenching substances: Sources, characteristics, and treatment. *Water Research*, 2018; **145**: 297–311.
31. Wiszniowski J, Robert · D, Surmacz-Gorska · J, Miksch · K, Weber J V. Landfill leachate treatment methods: A review. *Environ Chem Lett* 2006; **4**: 51–61.
32. Atabarut T, Ekinci E. Thermal Treatment of Landfill Leachate and the Emission Control. *Journal of Environmental Science and Health, Part A* 2007; **41**: 1931–42.
33. Zhao R, Xi B, Liu Y, Su J, Liu S. Economic potential of leachate evaporation by using landfill gas: A system dynamics approach. *Resources, Conservation and Recycling* 2017; **124**: 74–84.
34. Bai Z, Wang Y, Shan M *et al.* Study on anti-scaling of landfill leachate treated by evaporation method. *Water Science and Technology* 2021; **84**: 122–34.
35. Ye ZL, Hong Y, Pan S, Huang Z, Chen S, Wang W. Full-scale treatment of landfill leachate by using the mechanical vapor recompression combined with coagulation pretreatment. *Waste Management* 2017; **66**: 88–96.
36. Umar M, Aziz HA, Yusoff MS. Trends in the use of Fenton, electro-Fenton and photo-Fenton for the treatment of landfill leachate. *Waste Management*, 2010; **30**(11): 2113–2121.
37. Deng Y, Englehardt JD. Treatment of landfill leachate by the Fenton process. *Water Research*, 2006; **40**(20): 3683–3694.
38. Welter JB, Soares EV, Rotta EH, Seibert D. Bioassays and Zahn-Wellens test assessment on landfill leachate treated by photo-Fenton process. *Journal of Environmental Chemical Engineering* 2018; **6**: 1390–5.
39. Silva TFCV, Fonseca A, Saraiva I, Boaventura RAR, Vilar VJP. Scale-up and cost analysis of a photo-Fenton system for sanitary landfill leachate treatment. *Chemical Engineering Journal* 2016; **283**: 76–88.
40. Módenes AN, Espinoza-Quiñones FR, Manenti DR, Borba FH, Palácio SM, Colombo A. Performance evaluation of a photo-Fenton process applied to pollutant removal from textile effluents in a batch system. *Journal of Environmental Management* 2012; **104**: 1–8.
41. Primo O, Rivero MJ, Ortiz I. Photo-Fenton process as an efficient alternative to the treatment of landfill leachates. *Journal of Hazardous Materials* 2008; **153**: 834–42.
42. Kim SM, Vogelpohl A. Degradation of Organic Pollutants by the Photo-Fenton-Process. *Chemical Engineering and Technology* 1998; **21**: 187–91.
43. KIM Y-K, HUH I-R. Enhancing Biological Treatability of Landfill Leachate by Chemical Oxidation. *Environmental Engineering Science* 1997; **14**: 73–9.
44. Hermosilla D, Cortijo M, Huang CP. Optimizing the treatment of landfill leachate by conventional Fenton and photo-Fenton processes. *Science of the Total Environment* 2009; **407**: 3473–81.
45. Poblete R, Pérez N. Use of sawdust as pretreatment of photo-Fenton process in the depuration of landfill leachate. *Journal of Environmental Management* 2020; **253**: 109697.
46. Li L, Fu X, Ai J *et al.* Process parameters study and organic evolution of old landfill leachate treatment using photo-Fenton-like systems: Cu2+ vs Fe2+ as catalysts. *Separation and Purification Technology* 2019; **211**: 972–82.
47. Leszczyński J. Treatment of landfill leachate by using Fenton and photo-Fenton processes. *Journal of Ecological Engineering* 2018; **19**: 194–9.

48. Ghanbarzadeh Lak M, Sabour MR, Ghafari E, Amiri A. Energy consumption and relative efficiency improvement of Photo-Fenton – Optimization by RSM for landfill leachate treatment, a case study. *Waste Management* 2018; **79**: 58–70.
49. Huang D, Hu C, Zeng G *et al*. Combination of Fenton processes and biotreatment for wastewater treatment and soil remediation. *Science of the Total Environment*, 2017; **574**: 1599–1610.
50. Trapido M, Tenno T, Goi A *et al*. Bio-recalcitrant pollutants removal from wastewater with combination of the Fenton treatment and biological oxidation. *Journal of Water Process Engineering* 2017; **16**: 277–82.
51. Colombo A, Módenes AN, Góes Trigueros DE, Giordani da Costa SI, Borba FH, Espinoza-Quiñones FR. Treatment of sanitary landfill leachate by the combination of photo-Fenton and biological processes. *Journal of Cleaner Production* 2019; **214**: 145–53.
52. Silva JO, Silva VM, Cardoso VL, Machado AEH, Trovó AG. Treatment of sanitary landfill leachate by photo-Fenton process: Effect of the matrix composition. *Journal of the Brazilian Chemical Society* 2016; **27**: 2264–72.
53. Tejera J, Miranda R, Hermosilla D, Urra I, Negro C, Blanco Á. Treatment of a Mature Landfill Leachate: Comparison between Homogeneous and Heterogeneous Photo-Fenton with Different Pretreatments. *Water* 2019; **11**: 1849.
54. Li J, Zhao L, Qin L *et al*. Removal of refractory organics in nanofiltration concentrates of municipal solid waste leachate treatment plants by combined Fenton oxidative-coagulation with photo - Fenton processes. *Chemosphere* 2016; **146**: 442–9.
55. Silva TFCV, Ferreira R, Soares PA *et al*. Insights into solar photo-Fenton reaction parameters in the oxidation of a sanitary landfill leachate at lab-scale. *Journal of Environmental Management* 2015; **164**: 32–40.
56. Colombo A, Módenes AN, Trigueros DEG *et al*. Toxicity evaluation of the landfill leachate after treatment with photo-Fenton, biological and photo-Fenton followed by biological processes. *Journal of Environmental Science and Health - Part A Toxic/Hazardous Substances and Environmental Engineering* 2019; **54**: 269–76.
57. Singa PK, Isa MH, Ho YC, Lim JW. Mineralization of Hazardous Waste Landfill Leachate using Photo-Fenton Process. *E3S Web of Conferences*. Vol65. EDP Sciences, 2018: 05012.
58. De Morais JL, Zamora PP. Use of advanced oxidation processes to improve the biodegradability of mature landfill leachates. *Journal of Hazardous Materials* 2005; **123**: 181–6.
59. Jia CZ, Zhu JQ, Qin QY. Variation characteristics of different fractions of dissolved organic matter in landfill leachate during UV-TiO2 photocatalytic degradation. *Proceedings of the 2013 3rd International Conference on Intelligent System Design and Engineering Applications, ISDEA 2013*. 2013: 1594–8.
60. Cho SP, Hong SCSI, Hong SCSI. Photocatalytic degradation of the landfill leachate containing refractory matters and nitrogen compounds. *Applied Catalysis B: Environmental* 2002; **39**: 125–33.
61. Jia C, Wang Y, Zhang C, Qin Q. UV-TiO2 photocatalytic degradation of landfill leachate. *Water, Air, and Soil Pollution* 2011; **217**: 375–85.
62. Nomura Y, Fukahori S, Fujiwara T. Removal of 1,4-dioxane from landfill leachate by a rotating advanced oxidation contactor equipped with activated carbon/TiO2 composite sheets. *Journal of Hazardous Materials* 2020; **383**: 121005.
63. Teh CM, Mohamed AR. Roles of titanium dioxide and ion-doped titanium dioxide on photocatalytic degradation of organic pollutants (phenolic compounds and dyes) in aqueous solutions: A review. *Journal of Alloys and Compounds*, 2011; **509**(5): 1648–1660.

64. Elleuch L, Messaoud M, Djebali K *et al.* A new insight into highly contaminated landfill leachate treatment using Kefir grains pre-treatment combined with Ag-doped TiO2 photocatalytic process. *Journal of Hazardous Materials* 2020; **382**: 121119.
65. Shahmoradi B, Yavari S, Zandsalimi Y *et al.* Optimization of solar degradation efficiency of bio-composting leachate using Nd: ZnO nanoparticles. *Journal of Photochemistry and Photobiology A: Chemistry* 2018; **356**: 201–11.
66. Lien Thuong NT, Thanh Binh N, Small Scale Landfill Leachate Treatment Using Photocatalytic Oxidation Process 2015; **53**(3A): 49–54.
67. Azadi S, Karimi-Jashni A, Javadpour S, Amiri H. Photocatalytic treatment of landfill leachate using cascade photoreactor with immobilized W-C-codoped TiO2 nanoparticles. *Journal of Water Process Engineering* 2020; **36**: 101307.
68. Esfahani Kashitarash Z, Mohammad Taghi S, Kazem N, Abbass A, Alireza R. Application of iron nanaoparticles in landfill leachate treatment-case study: Hamadan landfill leachate. *Iranian Journal of Environmental Health Science and Engineering* 2012; **9**: 1–5.
69. Huang S, Li L, Zhu N *et al.* Removal and recovery of chloride ions in concentrated leachate by Bi(III) containing oxides quantum dots/two-dimensional flakes. *Journal of Hazardous Materials* 2020; **382**: 121041.
70. Ranjbari A, Mokhtarani N. Post treatment of composting leachate using ZnO nanoparticles immobilized on moving media. *Applied Catalysis B: Environmental* 2018; **220**: 211–21.
71. Mokhtarani N, Khodabakhshi S, Ayati B. Optimization of photocatalytic post-treatment of composting leachate using UV/TiO2. *Desalination and Water Treatment* 2016; **57**: 22232–43.
72. Chan AHC, Chan CK, Barford JP, Porter JF. Solar photocatalytic thin film cascade reactor for treatment of benzoic acid containing wastewater. *Water Research* 2003; **37**: 1125–35.
73. Meeroff DE, Bloetscher F, Reddy D V. *et al.* Application of photochemical technologies for treatment of landfill leachate. *Journal of Hazardous Materials* 2012; **209–210**: 299–307.
74. Cho SP, Hong SCSI, Hong SCSI. Study of the end point of photocatalytic degradation of landfill leachate containing refractory matter. *Chemical Engineering Journal* 2004; **98**: 245–53.
75. Poblete R, Otal E, Vilches LF, Vale J, Fernández-Pereira C. Photocatalytic degradation of humic acids and landfill leachate using a solid industrial by-product containing TiO2 and Fe. *Applied Catalysis B: Environmental* 2011; **102**: 172–9.
76. Saien J, Ojaghloo Z, Soleymani AR, Rasoulifard MH. Homogeneous and heterogeneous AOPs for rapid degradation of Triton X-100 in aqueous media via UV light, nano titania hydrogen peroxide and potassium persulfate. *Chemical Engineering Journal* 2011; **167**: 172–82.
77. Manassero A, Satuf ML, Alfano OM. Evaluation of UV and visible light activity of TiO2 catalysts for water remediation. *Chemical Engineering Journal* 2013; **225**: 378–86.
78. Karabelas AJ, Sarasidis VC, Patsios SI. The effect of UV radiant power on the rate of polysaccharide photocatalytic mineralization. *Chemical Engineering Journal* 2013; **229**: 484–91.
79. Jyothi KP, Yesodharan S, Yesodharan EP. Ultrasound (US), Ultraviolet light (UV) and combination (US+UV) assisted semiconductor catalysed degradation of organic pollutants in water: Oscillation in the concentration of hydrogen peroxide formed in situ. *Ultrasonics Sonochemistry* 2014; **21**: 1787–96.

134   **Photocatalytic Water and Wastewater Treatment**

80. Camera-Roda G, Santarelli F. Optimization of the thickness of a photocatalytic film on the basis of the effectiveness factor. *Catalysis Today* 2007; **129**: 161–8.
81. Rao NN, Chaturvedi V, Li Puma G. Novel pebble bed photocatalytic reactor for solar treatment of textile wastewater. *Chemical Engineering Journal* 2012; **184**: 90–7.
82. Yasmin C, Lobna E, Mouna M *et al.* New trend of Jebel Chakir landfill leachate pretreatment by photocatalytic TiO2/Ag nanocomposite prior to fermentation using Candida tropicalis strain. *International Biodeterioration and Biodegradation* 2020; **146**: 104829.
83. Ibrahim N, Selimin MT. Removal of Colour from Landfill by Solar Photocatalytic. *Journal of Applied Sciences* 2010; **10**(21): 2721–2724.
84. Hassan M, Wang X, Wang F, Wu D, Hussain A, Xie B. Coupling ARB-based biological and photochemical (UV/TiO2 and UV/S2O82−) techniques to deal with sanitary landfill leachate. *Waste Management* 2017; **63**: 292–8.
85. Vineetha MN, Matheswaran M, Sheeba KN. Photocatalytic colour and COD removal in the distillery effluent by solar radiation. *Solar Energy* 2013; **91**: 368–73.
86. Azadi S, Karimi-Jashni A, Javadpour S. Photocatalytic Treatment of Landfill Leachate Using W-Doped TiO2 Nanoparticles. *Journal of Environmental Engineering* 2017; **143**: 04017049.
87. Poblete R, Prieto-Rodríguez L, Oller I *et al.* Solar photocatalytic treatment of landfill leachate using a solid mineral by-product as a catalyst. *Chemosphere* 2012; **88**: 1090–6.
88. Azadi S, Karimi-Jashni A, Javadpour S. Modeling and optimization of photocatalytic treatment of landfill leachate using tungsten-doped TiO2 nano-photocatalysts: Application of artificial neural network and genetic algorithm. *Process Safety and Environmental Protection* 2018; **117**: 267–77.
89. Azadi S, Karimi-Jashni A, Javadpour S, Amiri H. Photocatalytic treatment of landfill leachate: A comparison between N-, P-, and N-P-type TiO2 nanoparticles. *Environmental Technology & Innovation* 2020; **19**: 100985.

doi: 10.2166/9781789061932_0135

Chapter 5
Life cycle assessment of solar photocatalytic wastewater treatment

Mohammadreza Hajbabaie[1], Hossein Nematollahi[2], Ka Leung Lam[3] and Alireza Bazargan[2*]

[1]Department of Civil Engineering, K.N. Toosi University of Technology, Tehran, Iran
[2]School of Environment, College of Engineering, University of Tehran, Tehran, Iran
[3]Department of Water Management, Faculty of Civil Engineering and Geosciences, Delft University of Technology, The Netherlands
*Corresponding author: alireza.bazargan@ut.ac.ir

ABSTRACT
This chapter conducts a life cycle assessment (LCA) of a solar-driven photocatalytic process for wastewater treatment. Initially, an overview is given of the LCA framework and how it has been used to assess the potential environmental impacts of different wastewater treatment technologies. The goal and scope of the LCA are defined, followed by the system boundary which includes all the major factors affecting the process, the treatment plant infrastructure, required chemicals, transportation, and electricity consumed during the processes. The foreground inventory data of the process are extracted from the literature, while the background inventory data are from the Ecoinvent3, ELCD and USLCI databases. Impact 2002+ has been used as the life cycle impact assessment method. In addition, the life cycle assessment of titanium dioxide (TiO_2), as one of most widely-used photocatalysts, is discussed. Finally, the key findings regarding the environmental impacts of solar-driven photocatalytic processes for wastewater treatment are addressed.

5.1 INTRODUCTION

Although technological advances have led to an increase in welfare, the planet is facing a series of big challenges. Climate change, source depletion, land use, water scarcity, pollution, and various diseases such as cancers can be traced back to human activity and its consequences. Improvements and modifications in consumption patterns are needed in order to reach sustainable development, meaning to meet the needs of today's generation, in addition to ensuring that our consumption patterns have the least possible negative impact on the lives of

© 2022 The Editors. This is an Open Access book chapter distributed under the terms of the Creative Commons Attribution Licence (CC BY-NC-ND 4.0), which permits copying and redistribution for noncommercial purposes with no derivatives, provided the original work is properly cited (https://creativecommons.org/licenses/by-nc-nd/4.0/). This does not affect the rights licensed or assigned from any third party in this book. The chapter is from the book Photocatalytic Water and Wastewater Treatment, Alireza Barzagan (Ed.).

136   Photocatalytic Water and Wastewater Treatment

future generations. Let us note that sustainability is not only an environmental issue, and should be considered in all aspects of economic, environmental, and social activities.

Investigating the environmental impacts of a product or service, from the extraction of the required raw material to the final disposal of the wastes, is essential in understanding its sustainability. The set of processes and stages from the production all the way to the final disposal of a product is called its life cycle, and the approach investigating the environmental impacts occurring in the life cycle is called life cycle assessment (LCA). In life cycle assessment, the system can be considered and examined from an economic, social or environmental perspective (or their combination). The term 'product system' is used which refers to all processes from the extraction of natural resources to the final management of waste produced at the end of the life cycle. This is also referred to as 'cradle to grave'. Cradle refers to the resource extraction stage and grave to the final waste disposal stage.

By using life cycle assessment, it is possible to choose the preferred scenario from several options, and to study and modify the emissions in the life cycle of a product, based on the impacts [1, 2]. For example, in selecting the best scenario between the use of plastic bottles and glass bottles for soft drinks, life cycle assessment can be carried out. At first glance, one may choose the glass bottle as the preferred scenario, assuming that avoiding the use of plastics is beneficial for the environment. However, by examining the life cycle of each of the two options, it is possible to scientifically study all the environmental effects that occur in each of the two scenarios, and make a more informed decision. Hence, according to the environmental impact categories considered, the preferred scenario with the least amount of effects can be selected. For the two options in this particular example, the following steps can be considered:

Plastic bottle

- Processes related to oil extraction
- Transfer of extracted oil to the factory to produce raw materials for plastic production
- Processes related to the production of plastic granules
- Transfer of the produced granules to the plastic bottle factory
- Processes related to making plastic bottles
- Transfer of plastic bottles to the soft drink factory
- Filling and packaging of the bottles
- Transferring of the filled bottles to stores
- Processes related to recycling and/or disposal of the discarded bottles

Glass bottle

- Extraction of silica and other required materials from mines
- Transfer of silica and other glass materials to the factory
- Processes related to material processing and glass bottle making
- Transfer of glass bottles to the soft drink factory
- Filling and packaging the bottles
- Transfer of the filled bottle to stores
- Processes related to recycling and disposal of used bottles

LCA of solar photocatalytic wastewater treatment

In each of these processes, energy and materials are consumed, and each process has a separate life cycle of its own. The energy consumed in each process can be in the form of electricity, heat, radiation and so on. The source of the electricity that is used is also important, meaning that the electricity can be generated from various renewable sources or fossil fuels, which make a significant difference in the environmental emissions and in turn affect the preferred option.

In essence, using the product system and life cycle approach, one can make a more accurate selection compared to when only a traditional view is taken. Often, the alternatives are numerous, for example the use of aluminum cans could also be added to the list of options above. For the particular example of food and beverage packaging, numerous academic papers have been published [3–5].

Overall, LCA provides a tool to quantify and evaluate the inputs, outputs, and potential environmental impacts of a product system over its life cycle, and its results may be used to support various decisions [6–8].

In recent years, despite many changes and developments, life cycle assessment has been standardized and attempts have been made to minimize discrepancies. For example, as per the ISO standards, the four basic steps of an LCA include the following which will be subsequently discussed:

- Goal and scope
- Life cycle inventory
- Life cycle impact assessment
- Interpretation

5.2 THE BASIC STEPS OF LCA

5.2.1 Goal and scope

The first step in a life cycle assessment is to define the goal and scope of the work. Defining the goal and scope of the LCA study unambiguously and at this stage, before any data collection and calculations are performed, is important to avoid possible problems down the road.

The purpose of a life cycle assessment can be any of the following:

- Selecting a preferred alternative
- Optimizing the life cycle of a product to minimize impacts
- Providing energy and environmental labels
- Marketing

At this stage, the purpose of the study, the options that will be compared with each other, how to use the study results, the data collection method, the system under study, system boundaries and performance units are determined. The product, process, or activity is also defined and the system hypotheses are documented and collected. This stage provides the necessary basis for the LCA. The scope expresses the framework in which the study is conducted and should be consistent with the objectives of the assessment. Reliable information will not be obtained if the goal and scope of the assessment are not specified.

138 Photocatalytic Water and Wastewater Treatment

The 'system boundary' is defined as the range within which all the processes in a life cycle are located according to the intended purpose and accuracy of the LCA. In other words, the system boundary determines which unit processes to include in the LCA study [9]. Within the system boundary, all inputs and outputs, including material flow, energy and emissions, are included. The accuracy of a life cycle assessment study depends on the accuracy of the system boundary selection. The more accurate the system boundary is, the higher the accuracy of the results. Lack of sufficient reliable data of included processes and the complexity level of the model are the obstacles to consider when designating system boundaries.

In addition, defining a 'functional unit (FU)' is essential for building and modeling a product system. The purpose of defining a functional unit is to provide a quantified reference unit for normalizing the inventory. The definition of a functional unit depends on the type of environmental impacts and the purpose of the study. The functional unit is often defined in terms of the mass, number and volume of the product produced in the study. Due to the linear nature of LCA calculations, the results can also be matched to reality by multiplying the ratio of the manufactured product to the functional unit in the final results.

To put the importance of the FU into perspective, one can consider this example: in comparing two different types of batteries in an LCA, shall the researcher choose a specific number of batteries as the FU with which the alternatives are compared? Or would it be better to compare the results based on a FU of 'Watt hours of energy provided'? Perhaps in comparing battery A and battery B, battery A might provide 50% more energy than battery B, but have 20% more environmental impacts as well. In such a case, choosing an FU of the number of batteries would hint at battery B being a better option because of its relatively lower environmental impact, but if the amount of energy provided is chosen as the FU, then battery A would be the better choice. Evidently, in this simple case, as in many other real-world cases, the choice of the FU could have drastic influence on the final results [10, 11].

5.2.2 Life cycle inventory

The second step to consider in conducting LCAs is the life cycle inventory, which involves collecting the required data based on a list of input and output materials, emissions and energy. Life cycle inventory includes the collection and calculation of inventory and quantification of inputs and outputs for a product throughout its life cycle [12].

Since this step only involves collecting input and output data and data analysis, it is not possible to achieve the appropriate conclusions from this step alone without performing the next step, which is to determine the environmental impacts of processes within the life cycle.

The inventory within the system boundary can be classified under the following major headings:

- Energy, raw materials, and emission inputs
- Products, by-products, and final wastes

LCA of solar photocatalytic wastewater treatment 139

- Emissions to the air, water and soil
- Other environmental aspects

After collecting the required inventory, calculation methods include:

- Validation and verification of the collected data
- Determining the relationship between inventory and unit processes
- Determining the relationship between data and base flow in the functional unit

This phase of life cycle assessment is more time consuming than other phases. This is often due to the time spent on collecting information. Data are collected in a short period when there are good baseline data. Also, if the suppliers of materials to the process cooperate with the person(s) carrying out the LCA, data collection can be greatly improved and streamlined. Much of the basic information needed is available and can usually be gathered in software designed for this purpose. Data can be collected based on transportation, extraction of raw materials, processing of materials, and so on, and are often provided in the software.

However, it should be noted that although a number of software packages are available to help conduct LCA studies in line with the ISO 14040/14044 standards, systematic comparisons have shown that the results of the LCA study are unfortunately highly dependent on which software is used. An evaluation by Speck *et al.* found that various programs disagreed over which packaging containers had the greatest environmental impact and that their results in some cases were over an order of magnitude off. In all categories of impact and software investigated, the results of the analysis were different from each other at multiple points in the comparisons [13, 14]. Such stark differences in results depending on the software which is used have also been reported elsewhere, the implications of which have been referred to as 'worrying', and rightfully so [15]. Even when exactly identical inputs are used, the results may turn out to be different depending on which software is used.

5.2.3 Life cycle impact assessment

Life cycle impact assessment (LCIA) helps to understand and assess environmental impacts based on inventory analysis in the context of the goal and scope of the study. The impact assessment phase includes the selection of impact categories, and classification and description of environmental impacts. This step also provides the necessary information for the interpretation phase.

The basic structure of impact assessment methods is based on the following five principles [16]:

1. Selection
2. Characterization
3. Damage assessment
4. Normalization
5. Weighting or grouping

140 **Photocatalytic Water and Wastewater Treatment**

Items 3 to 5 are not present in all impact assessment methods and are optional elements [16]. When choosing an impact assessment method, the required items can be selected to be applied to the assessment method.

5.2.3.1 Selection

Choosing impact categories should reflect the parameters that were selected for the assessment as part of the scope definition. Each impact category is then assigned a representative indicator along with an environmental model that can be used to calculate how elementary flows affect the indicator.

5.2.3.2 Characterization

At this stage, life cycle inventory (LCI) outcomes are assigned to selected impact categories based on known environmental impacts. In practice, this is often complemented by the use of LCI databases or LCA software. This means the amount of materials that are classified as pollutants in groups affecting the environment is multiplied by a value called the characterization factor to indicate the degree of participation and impact of this material in determining the impacts on the environment. In each impact category, a common criterion is used to calculate the scores. In this way, each contribution is combined into a single score that represents the impact of each category on the product system as a whole. For example, using the global warming potential (GWP), one can estimate how much energy a ton of each gas absorbs over a given time period, such as 100 years, compared with a ton of carbon dioxide. In comparison to carbon dioxide, methane absorbs a great deal more energy; hence its GWP varies from 28 to 36 [17].

5.2.3.3 Damage assessment

In some methods of impact assessment, there is a step called damage assessment. In this step, the impact of classification indicators that have a common unit are added. For example, in the Eco-Indicator99 method, all effects related to human health are expressed with DALY (disability adjusted life years). For example, by using DALY, it is possible to summarize the effects of all carcinogens and other human health threats at the same time [18].

5.2.3.4 Normalization

In many impact assessment methods, it is possible to compare the results of different impact assessment indicators with each other based on a certain criterion or target value; this target value is used as a comparison reference. Normalization factors show the impact of an entire reference area for a particular type of impact (e.g., climate change, eutrophication, etc.) in a reference year. Based on the selected impact assessment method, the normalization results can be used in the stages of characterization and damage assessment.

5.2.3.5 Weighting or grouping

Weighting is the application of quantitative measures to the relative severity of various environmental changes. In some impact assessment methods, it is possible to weight the environmental impact classification. This means that the results obtained from the impact classification indicators are multiplied by the weighting factor and create a final score for the amount of impact. Weighting can be done on normalized or non-normalized data.

5.2.4 Interpretation

Interpretation is the mandatory final phase of an LCA that links inventory analysis and impact assessment to obtain robust conclusions and recommendations. The interpretation should reflect the fact that the results of the impact assessment are based on a relative and comparative approach and express only the environmental impact potential. It should also note the fact that the results do not have the power to predict the threshold and the risk limits of the effects.

At this phase, the results concerning the goal and scope of the study will be interpreted and management strategies will be conferred. The interpretation of the results can be presented to support decision-makers as recommendations. Interpretation is a systematic method for identifying, controlling and assessing information from the results of inventory analysis.

Interpretation involves considering all phases of the assessment and examining all consistent hypotheses. Ultimately, there are three elements to the LCA interpretation phase as follows [19]:

(1) Identification of significant issues (based on the results of the LCI and LCIA phases)
(2) Evaluation that considers completeness, consistency and sensitivity checks
(3) Conclusions, limitations and recommendations

5.3 LCA FOR WASTEWATER TREATMENT AND PHOTOCATALYSIS

In most LCA studies for wastewater treatment, the goal is to assess the environmental impacts of technologies. System boundaries are often chosen so that all inventories are considered during the process. Figure 5.1 shows an

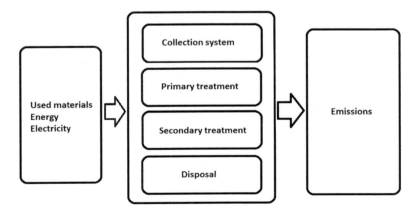

Figure 5.1 Examples of physical system boundaries for wastewater treatment plant LCAs.

example of the system boundary for an LCA study for wastewater treatment. Generally, 1 m^3 of input flow is used as the functional unit.

System inputs are usually the wastewater entering the system, the electrical energy used for pumping and mixing, aeration and the chemicals used. Outputs include emissions to air, water, soil, and other secondary products such as sludge.

Over the past couple of decades, there has been a great deal of interest in using LCAs in the field of wastewater treatment [20–22]. Studies have summarized past LCAs and describe key challenges in using LCAs in wastewater systems [23], and using LCA to evaluate sludge treatment and disposal [24]. One study examined the use of LCAs for a broader description of urban water systems [25]. Wastewater treatment plants can come in a wide range of shapes and sizes. For some, the operation phase is most important while for others the construction and decommissioning are also significant. For example, it is important to include the construction phase in the LCA of low complexity treatment systems in developing countries [26], which often suffer from a lack of site-specific characterization factors and normalization/weighting data [27].

According to one study, for conventional plants, the environmental impacts of the construction and ultimate deconstruction phase of a wastewater treatment plant are negligible compared to the operation phase [28]. Different wastewater treatment processes were investigated to determine the use of materials and energy as well as their environmental effects. According to the results of this study, the energy used in the operation phase of the treatment plant had a large contribution to the environmental effects of the treatment process. Also, biological filters consume less energy and their emissions are less than those of the activated sludge process.

Suh and Rousseaux [29] compared different scenarios for the disposal of sludge from wastewater treatment plants. The main scenarios were burning, using in agricultural land or disposal in the ground. The sludge stabilization process was chosen among anaerobic mesophilic digestion, using directly as fertilizer in agriculture, and lime stabilization. The results of this study are weighted and normalized and according to the findings, using sludge in agriculture has the lowest energy consumption however this option, alongside burning, releases heavy metals which have the maximum contribution in human and ecological toxicity.

Dixon *et al.* [30] studied two systems for wastewater treatment in terms of environmental impacts containing a reedbed and biofilters. The studied impacts include energy use, released carbon dioxide and solid emissions. Based on the results of this study, the reedbed system has less CO_2 emission and energy consumption compared to biofilters, but in terms of solid emissions, environmentally it is less desirable than biofilters. Most of the energy consumed in biofilters was due to the aeration and pumping system during operation. Most of the solids emissions of the reedbed system were due to drilling in the construction phase and the sludge of the operation phase [30].

Renzoni *et al.* [31] conducted LCA starting from the pumping station all the way to the wastewater treatment plant. According to the results of their study, the greater the volume of water treated, the lower the eutrophication and acidification, while the environmental impacts in the other categories such as climate change and toxicity increase. Using normalization, acidification and eutrophication were determined to be the most important impacts and therefore, increasing the rate of wastewater treatment as much as possible has been suggested. Meanwhile, Foley *et al.* [32] stated that advanced nitrogen removal processes require the addition of complementary chemicals, and that the negative environmental effects associated with the transportation and manufacture of these materials is usually ignored.

The LCA of photocatalytic systems has also been explored in the literature [33]. In one study, a comparative LCA of two solar photocatalytic processes (heterogeneous photocatalysis and homogeneous photo-Fenton) coupled with biological treatment was carried out [34]. The study used α-methyl-phenylglycine as a target substance. The system boundary included the construction phase, chemicals, electricity, transport, end of life of the spent catalyst, sludge incineration, as well as treatment of the effluent. Nine environmental impact categories, namely ozone depletion, freshwater aquatic toxicity, human toxicity, global warming, photochemical ozone formation, eutrophication, energy consumption, acidification, and land use were investigated. The study suggests that if a solar photo-Fenton process is used in an industrial application, less time and energy is needed for obtaining biodegradable effluents than when solar heterogeneous photocatalysis is performed. In other words, the solar heterogeneous photocatalytic system displays a much higher environmental impact in nearly all categories, mainly due to the larger size of its solar collector and the land required [34].

For olive mill wastewater treatment, when comparing ultraviolet (UV) heterogenous photocatalysis (UV/TiO$_2$), wet air oxidation and electrochemical oxidation over boron-doped diamond electrodes, the photocatalytic process did not appear very attractive in terms of environmental performance. It was also concluded that human health is primarily affected followed by impacts on resources depletion, and that energy requirements strongly impact the sustainability of the process [35].

Real life cycle inventories were gathered in another study which considered various forms of treatment, namely: solar photolysis (both simulated and real), photolysis under different UV irradiations (UV-A and UV-C), solar photo-Fenton in the absence or presence of hydrogen peroxide (denoted solar/Fe and solar/Fe/H$_2$O$_2$ respectively), titania-mediated photocatalysis (UV-A/TiO$_2$), and UV-C treatment in the presence of hydrogen peroxide (UV-C/H$_2$O$_2$). The functional unit for comparing the various considered forms of treatment was the removal of 1 μg of 17α-ethynylestradiol from one liter of wastewater. The chosen pollutant is an endocrine disrupting chemical commonly found in micro concentrations. Solar photolysis alone, without Fe, showed a 23-fold increase in environmental footprint. If hydrogen peroxide was used in addition to Fe, the environmental footprint further decreased. Also, the use of simulated solar irradiation significantly increased the environmental footprint due to its energy intensive nature. UV-C was found to be about three times more environmentally friendly than UV-A photolysis. Addition of TiO$_2$ to UV-A and H$_2$O$_2$ to UV-C caused their total environmental impacts to significantly decrease meaning that the excess burden of using these materials was more than compensated for by their improvement of process efficiency. The total environmental footprint could be

144   **Photocatalytic Water and Wastewater Treatment**

ordered from worst to best as: solar photolysis >UV-A > UV-C > solar/Fe > UV-A/ TiO_2 > UV-C/H_2O_2 > solar/Fe/H_2O_2 [36]. Similarly, another study also compared numerous solar-driven processes including solar photolysis (with and without H_2O_2), photocatalysis using TiO_2 (with and without H_2O_2) and circumneutral photo-Fenton [7]. This study concluded that the use of titanium dioxide alongside solar irradiation provided the best balance in terms of minimizing environmental impacts while achieving acceptable removal of pollutants. Elsewhere [8], using 1% of manganese, iron, nickel or cobalt to monodope TiO_2 for producing an improved photocatalyst for removing carbamazepine and methyl orange was studied. The research showed that although photocatalytic activity of the catalyst was greatly influenced by the use of the metals, the environmental impacts of synthesizing the TiO_2 did not significantly change when a weight-based functional unit was employed. This was true irrespective of using UV-A or visible light irradiation. The EATOS (Environmental Assessment Tool for Organic Synthesis) has been used alongside LCA to study the effects of functionalization [37].

Specifically comparing UV-A lamps and solar energy, Muñoz *et al.* [38] investigated heterogenous photocatalysis, photo-Fenton reactions, the coupling of heterogeneous photocatalysis and photo-Fenton, and heterogeneous photocatalysis in combination with hydrogen peroxide for kraft mill bleaching wastewater. The study concluded that no single process was better than the rest in all impact categories, however the environmental impacts of all the advanced oxidation processes (AOPs) under study are caused mainly by the amount of electricity consumed, and that the impact of producing the reagents and catalysts is comparatively low. In a follow up study, the same team conducted a similar study but this time included ozonation alongside the other treatments for the removal of dissolved organic content [39].

Researchers recently used LCA to assess the impacts of treating $100\,m^3$/d of wastewater containing $100\,mg/L$ of phenol using the baseline CML impact factors. The single score Eco-indicator 99 was used alongside other individual impact categories. Overall, the solar photo-Fenton process received the lowest Eco-indicator score of 0.044 pt making it the most environmentally friendly. Meanwhile, the electro-Fenton process obtained the highest Eco-indicator score of 1.48 pt making it the least favorable option. Solar photocatalysis using various catalysts as well as adsorption by activated carbon ranked between the abovementioned technologies [1].

Overall, the benefits of LCA for water and wastewater managers can be widespread. The information learned through proper conduction of LCA can cover various aspects of the entire wastewater system. Whether upstream issues and components such as supply chain are of concern, or whether downstream issues such as waste disposal or point-of-use are of interest, LCA can be a useful tool. Other benefits could be the identification of weaknesses and bottlenecks for targeted improvement and optimization. Another important benefit of LCA can be the identification of trade-offs and burden-shifting [20].

5.4 MATERIALS AND METHODS OF THE CURRENT STUDY

As described previously, life cycle assessment consists of four main stages [40, 41]. The first stage in conducting a cradle-to-grave LCA study is goal and scope

LCA of solar photocatalytic wastewater treatment 145

definition. Defining a functional unit and the boundaries of the analysis are conducted in this phase. The second stage is the life cycle inventory, in which the modeling of the system is done considering the input or output flows. The third step is to assess the environmental impacts. At this stage, all inventory emissions are classified into one or more designated damage categories such as carcinogenicity, environmental toxicity, or fossil fuel use according to their damage potential. The last stage is interpretation. This step sometimes involves weighting the impact categories according to human preferences to achieve a single impact score that can be compared for alternative products or processes.

5.4.1 Goal and scope of the current study

The goal of this study is to assess the environmental impacts related to photocatalytic water and wastewater treatment using life cycle assessment of a particular case study. The boundary of the system is illustrated in Figure 5.2. Importantly, titanium dioxide is considered as the only photocatalytic material used herein, and it is modeled according to the article 'Life cycle of titanium dioxide nanoparticle production: impact of emissions and use of resources' [42]. If a different photocatalytic material or process were used, this would strongly impact the results of the current work. The solar plant for solar-driven advanced oxidation processes, the area occupied, electricity, transportation and raw materials are included in the inventory. The system boundary for the LCA is shown and in the modeling section, titanium dioxide is modeled separately and entered into the system. As discussed previously, the functional unit is one of the main elements in the definition of goal and scope. In this study, the selected functional unit is to treat 1 m^3 of wastewater to eliminate toxic and non-biodegradable compounds to achieve pollution levels in the effluent which are acceptable for wastewater discharge. The characteristics of the wastewater inlet and outlet are given in Table 5.1.

5.4.2 Inventory analysis of the current study

The study was conducted at the life cycle scale, containing all energy, materials and fuel inputs to the process and their upstream emissions. The process inventories in the software are a black box, meaning that the processes described by the input and output streams are presented without any further information about the internal functional relationships. For titanium dioxide, a cradle-to-gate life cycle assessment has been used, and the final product of that process (titanium dioxide) is used in the photocatalytic process under investigation.

Ecoinvent3, ELCD and USLCI databases have been used to collect data. Table 5.1 shows the input and output characteristics of the wastewater. Table 5.2 indicates the inventories according to the defined system boundary. Table 5.3 shows the inventory for modeling titanium dioxide.

5.4.3 Life cycle impact assessment of the current study

Among the many methods available in the SimaPro software, IMPACT 2002+ has been used as the life cycle impact assessment method in this project. Initially developed in 2003, the new IMPACT 2002+ life cycle impact assessment method proposes the practical implementation of a hybrid midpoint/damage

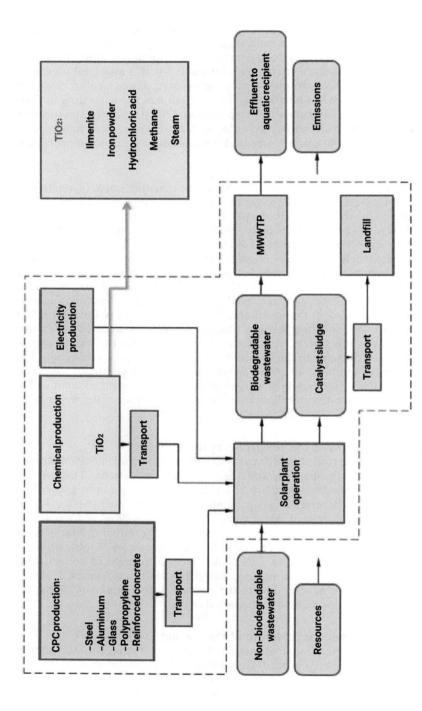

Figure 5.2 Flow diagram and system boundaries in the LCA study. MWWTP, municipal wastewater treatment plant; CPC, compound parabolic collector.

LCA of solar photocatalytic wastewater treatment

Table 5.1 Initial and final effluent characteristics [34].

Initial Effluent (Non-Biodegradable)

a-methyl-phenylglycine (MPG) (mg/L)	500
DOC (mg/L)	330
COD (mg/L)	1270
N-total (mg/L)	42
N-ammonia (mg/L)	0
N-nitrate (mg/L)	0
Final Effluent (Biodegradable)	
MPG (mg/L)	0
DOC (mg/L)	40
COD (mg/L)	214
N-ammonia (mg/L)	36
N-nitrate (mg/L)	<1
Accumulated UV Radiation and Time Required	
Q_{uv} (kJ/L)	252
$t_{30\,W}$ (min)	1500

approach, linking different types of life cycle inventories (primary streams and other interventions) through 14 midpoints and 4 damage categories. Midpoint categories can be used to identify traditional impact assessment methods. Also, the uncertainties that occur due to the cause and effect chain can be restricted by using midpoint categories. Damage categories are also called endpoint categories. Each damage category is obtained by aggregating the effects of the midpoint categories. Four main endpoint categories, namely: resource use, human health damage, climate change, and ecosystem quality are used. The endpoint category regarding human health is split into carcinogenic and non-carcinogenic damages. Regardless of whether toxicity in its broad sense or toxicity only for humans is considered, main responses are used rather than assumptions. Existing methods such as Eco-indicator 99 and CML 2002 are used for other midpoint categories, the scores of which are expressed in reference material units [43].

The results can (and should) be normalized. This can occur at midpoints or endpoints as needed. The characteristic factors of more than 1000 different life cycle inventory outcomes are now available when using methods such as the IMPACT 2002+ [43]. Each end point consists of one or more midpoints shown in Table 5.4 alongside their respective units.

The DALY is a unit for measuring adverse effects on human health; 1 DALY is the absence of one year of life (e.g., due to dying one year too soon). Importantly, the number of DALYs can be converted depending on the hardship or disability inflicted. For instance, a person suffering with 25% disability for 4 years also incurs 1 DALY. The unit for ecosystem quality, 1 PDF.m²yr, means that all species disappear from 1 m² of land in one year; which is also equivalent to 10% of all species disappearing from 10 m² of space in one year; or 10% of all species disappearing from 1 m² of land over 10 years. In the case of climate

148 **Photocatalytic Water and Wastewater Treatment**

Table 5.2 Inventory table for 1 m³ wastewater treatment [34].

Inputs (per Functional Unit)	1 m² Compound Parabolic Collector	Solar Heterogeneous Photocatalysis
From nature		
Occupation, industrial site (m² year)	2.76	2.4
From technosphere CPC infrastructure:		
Stainless steel (kg)	7.81	0.45
Galvanized steel (kg)	0.17	9.5×10^{-3}
Aluminum extruded and anodized * 2 (kg)	9.68	0.55
Borosilicate glass tube * 2 (kg)	6.72	0.38
Extruded polypropylene (kg)	0.2	1.1×10^{-2}
Injection molded polypropylene (kg)	1.2	0.069
Concrete (m³)	0.32	0.018
Reinforcing steel (kg)	31	1.8
Concrete blocks (m³)	0.02	1.1×10^{-3}
Materials transport by rail (kg.km)	14,089	807
Materials transport by lorry 32 t (kg km)	45,579	327
Auxiliary materials and energy:		
Electricity, medium voltage, UCTE profile (kWh)		18
Titanium dioxide (kg)		0.02
Chemicals transport by rail (kg km)		12
Chemicals transport by lorry 32 t (kg km)		2
Outputs (per functional unit) To nature		
Carbon dioxide (kg)		1.06
To technosphere Spent catalyst management:		
Transport by lorry 16 t (kg km)		20
Catalyst landfilling (fresh weight) (kg)		0.40
Effluent treatment in MWWTP (m³)		1

Table 5.3 Inventory for titanium dioxide production [42].

Inputs	Mass (kg/kg TiO_2)	Energy (MJ)	Exergy (MJ)
Ilmenite	2.165	0	1.928
Iron powder	0.103	0	0.691
Hydrochloric acid	0.065	0	0.151
Methane	0.866	44.894	46.690
Steam	14.948	2.559	4.572
Electricity	-	5.443	5.443

LCA of solar photocatalytic wastewater treatment   149

Table 5.4 Midpoints and endpoints in the IMPACT 2002+ method alongside their respective units [40, 44–48].

Midpoint Category	Midpoint Reference Substances	Damage Category (Endpoints)	Damage Unit
Human toxicity (carcinogens + non-carcinogens)	kg chloroethylene into air	Human health	DALY
Respiratory (inorganics)	kg $PM_{2.5}$ into air	Human health	
Ionizing radiation	Bq_{eq} carbon–14 into air	Human health	
Ozone layer depletion	kg_{eq} CFC–11 into air	Human health	
Photochemical oxidation [=Respiratory (organics) for human health]	kg ethylene into air	Human health / Ecosystem quality	–
Aquatic ecotoxicity	kg_{eq} triethylene glycol into water	Ecosystem quality	PDF.m^2yr
Terrestrial ecotoxicity	kg_{eq} triethylene glycol into water	Ecosystem quality	
Terrestrial acidification/ nutrification	kg_{eq} SO_2 into air	Ecosystem quality	
Aquatic acidification	kg_{eq} SO_2 into air	Ecosystem quality	Under development
Aquatic eutrophication	kg PO_4^{3-} into water	Ecosystem quality	Under development
Land occupation	m^2_{eq} organic arable land-year	Ecosystem quality	PDF.m^2yr
Global warming	kg_{eq} CO_2 into air	Climate change (life support system)	($kg_{eq}CO_2$ into air)
Non-renewable energy	MJ total primary non-renewable or kg_{eq} crude oil (860 kg/m^3)	Resources	MJ
Mineral extraction	MJ additional energy or kg_{eq} iron (in ore)	Resources	

change, the main issue is global warming potential of emissions, and this value is obtained by calculating the amount of carbon dioxide equivalent produced. Finally, the unit used for resources is kg.

5.5 RESULTS AND DISCUSSION

In interpreting the results of the LCA, the approach and vision of the researcher is very important. In order to be more objective, the analysis should be based on the goal and scope, defined in the first stage of an LCA study.

From the results of the LCA carried out in SimaPro software, it is evident that in all midpoint categories, transportation plays a key role in the emission of pollutants. The transportation process includes the transportation of fuels required for bringing raw materials to the required location, as well as the processes associated with vehicle production (which is itself a separate life cycle assessment project). Thus, a significant reduction of the environmental burden could be achieved by reducing the distance required for transporting the raw materials. This objective can be met by purchasing raw materials from more local production centers.

Following the transportation process, raw materials such as concrete and aluminum have the most destructive environmental effects. As discussed earlier, each of the processes in a life cycle can be considered separately and in more detail as a life cycle assessment project. In a software database, many of these processes are calculated for a specific functional unit, and the environmental impacts of each can be obtained using an existing database. Regarding the processes related to the production of concrete, aluminum and steel, it should be noted that all the calculations from the extraction of their raw materials to the processes required for the production of the product have been previously studied.

Cement production is associated with a significant environmental burden and is a major CO_2 emitter due to the clinker process and fuel combustion. Extraction, clinker and transportation emissions are the main processes that contribute to the concrete production process [49]. The main processes related to the production of aluminum are extraction, the conversion of bauxite to alumina, the processing of alumina to aluminum, the final cast products and transportation processes. In the reduction process, molten salt electrolysis is used to produce aluminum melt. Direct electrical current is required for electrolysis, which is provided by rectifiers. Under the influence of this direct current, aluminum oxide (alumina) is converted to pure aluminum [50]. For the steel production process, extraction processes, metallurgical processes and transportation lead to environmental emissions. Metallurgical processes in the steel industry are very energy-consuming and the production of crude steel will emit significant amounts of CO_2 [51]. Each of these processes requires the consumption of energy and raw materials, which leads to the production of environmental pollutants. Subsequently, each of the pollutants with a certain coefficient is involved in each of the impact categories.

As can be seen in Figure 5.3, the consumption of fossil fuels in each of these processes leads to CO_2 emissions, which is one of the most important factors in global warming [52]. In this study, the most important contribution to global warming comes from transportation, followed by cement production. Also from Figure 5.3, it can be discerned that due to the consumption of raw materials for the production of concrete, steel and aluminum, as well as the raw materials required for the production of vehicles, these four processes are more influential in the resource extraction category.

The use of coke is common in the steel industry. Consumption of coke leads to the production of benzo(a)pyrene (BaP), a mutagenic and highly carcinogenic polycyclic, and therefore the steel process plays a major role in the category of

LCA of solar photocatalytic wastewater treatment

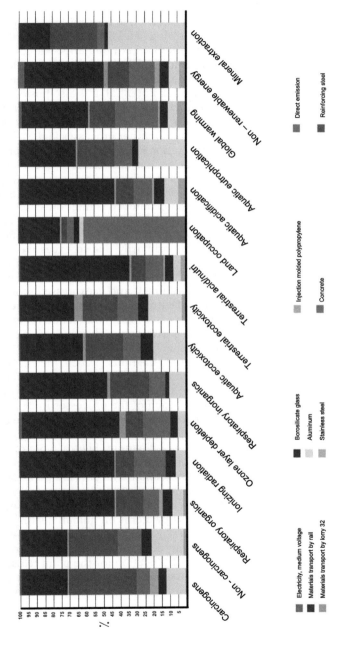

Figure 5.3 Environmental impacts for the functional unit of 1 m³ of wastewater treatment with solar photocatalysis.

carcinogenic impact [53]. The effects of air pollution released during transportation can lead to respiratory and cardiovascular effects. For this reason, in the category of respiratory effects, the highest impact is related to transportation, which leads to the release of pollutants due to the burning of fossil fuels.

SO_2, NOx and NHx are the most important acidic pollutants. The combination of these gases with water can lead to acid rain or the direct acidification of bodies of water. The emission of these pollutants from the combustion of fossil fuels is an important source of acidification potential. It is also possible to emit these pollutants in the processes related to the production of concrete, steel and aluminum. Therefore, in the process of terrestrial acidification, these four processes have a more significant effect than other processes [54].

For ozone depletion, again, transportation is the main culprit. As previously reported in the literature, transportation has the greatest effect on ozone depletion through the release of halogens and chlorofluorocarbons (CFCs) and the photochemical oxidation resulting from the use of fossil fuels [55]. Similarly, pollutants released during fuel extraction and consumption can enter the aquatic environment and lead to pollution of water resources and that is why transportation plays a major role in the aquatic ecotoxicity category [56].

Eutrophication of aquatic systems is primarily due to excessive intake of nitrogen and phosphorus (mostly due to excessive use and runoff of fertilizer in the real world). In LCA, eutrophication potential is a measure of emissions that cause eutrophic effects in the environment and is expressed in kilograms equivalent to phosphate. The potential of eutrophication (mainly NOx emission) in the current study comes from the processes of aluminum and reinforcement steel (rebar) production as well as transportation which has a major impact on the classification of aquatic eutrophication [57].

Figure 5.4 shows damage assessment diagrams, which include the four endpoint categories: human health, ecosystem quality, climate change and resources. This chart relatively examines each of the endpoint categories on a percentage scale. Given that the sum of each of the processes in the endpoint classification is 100%, with such a diagram, comparisons can be made between the processes in a particular endpoint category, but not between endpoint categories. The normalization diagrams shown in Figure 5.5 are used to compare the endpoint categories relative to each other.

As explained in the impact assessment section, using the midpoint categories and their aggregation using specific coefficients, the endpoint categories can be calculated. According to Figure 5.5, it can be seen that the greatest impact of the solar photocatalytic treatment of wastewater is on human health. It should be kept in mind that these conclusions are only true regarding the defined goal and scope. If a different photocatalytic process, different photocatalytic material, different wastewater, or different location for the plant (closer to the raw materials) was chosen, then results could have been different [58]. On the other hand, it can be seen to what extent each process affects each of the endpoint categories in Figure 5.5.

According to Figure 5.6, all processes are divided based on four endpoint categories: human health, ecosystem quality, climate change and resources. As can be expected, transportation has the greatest impact.

LCA of solar photocatalytic wastewater treatment 153

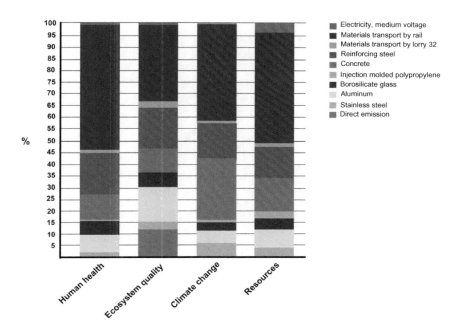

Figure 5.4 Damage assessment diagram for 1 m³ of solar photocatalytic wastewater treatment using the SimaPro software with the IMPACT 2002+ method.

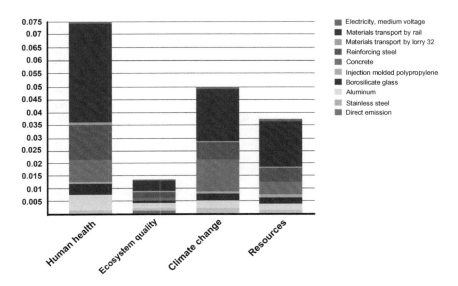

Figure 5.5 Normalization diagram for 1 m³ of solar photocatalytic wastewater treatment using the SimaPro software with the IMPACT 2002+ method.

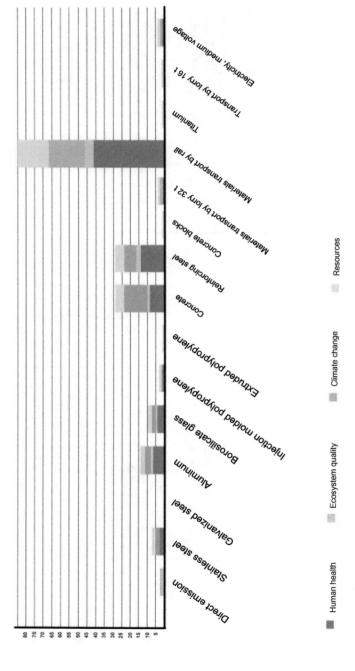

Figure 5.6 Single score chart for 1 m³ of solar photocatalytic wastewater treatment using the SimaPro software with the IMPACT 2002+ method.

It should be noted that the life cycle assessment process is a complex process with black-box inventories and the more details which are observed in the system demarcation and logging, the more accurate the results will be. For instance, the amount of carbon dioxide produced by the labor force while working can be considered as part of the emissions to air. It is also possible to consider water consumption or raw materials required for manufacturing the treatment equipment as well as all the relevant emissions that enter the air, water and soil in the inventory to increase the accuracy of the study.

5.6 CONCLUSION

In this chapter, life cycle assessment has been used as an approach to estimate the environmental damage of treating 1 m³ of wastewater with solar photocatalytic treatment. Titanium dioxide was chosen as the photocatalytic material. According to the study conducted on the treatment of 1 m³ of municipal wastewater using solar heterogeneous photocatalysis, it is observed that the main factors in the emission of pollutants into the environment are transport, reinforcing steel and concrete. The take-home message from this study is that by reducing the requirements for transporting raw materials and/or by changing the production method of the materials, the emissions can be reduced.

REFERENCES

1. Magdy M, Alalm MG, El-Etriby HK. Comparative life cycle assessment of five chemical methods for removal of phenol and its transformation products. *Journal of Cleaner Production* 2021; **291**: 125923.
2. Hossain MU, Guo M-Z, Poon CS. Comprehensive environmental evaluation of photocatalytic eco-blocks produced with recycled materials. In: de Brito J, Thomas C, Medina C, Agrela F, eds. *Waste and Byproducts in Cement-Based Materials*. Elsevier, 2021: 567–82.
3. Boesen S, Bey N, Niero M. Environmental sustainability of liquid food packaging: Is there a gap between Danish consumers' perception and learnings from life cycle assessment? *Journal of Cleaner Production* 2019; **210**: 1193–206.
4. Verghese K, Lockrey S, Clune S, Sivaraman D. 19 - Life cycle assessment (LCA) of food and beverage packaging. In: Yam KL, Lee DS, eds. *Emerging Food Packaging Technologies Principles and Practice*. Woodhead Publishing, Sawston, UK, 2012: 380–408.
5. Saleh Y. Comparative life cycle assessment of beverages packages in Palestine. *Journal of Cleaner Production* 2016; **131**: 28–42.
6. Espino M, Gomez FJ V, Boiteux J, Fernández M de los Á, Silva MF. 2.53 - Green Chemistry Metrics. In: Comprehensive Foodomics. Cifuentes A, eds. Oxford: Elsevier, 2021: 825–33.
7. Pesqueira JFJR, Pereira MFR, Silva AMT. A life cycle assessment of solar-based treatments (H2O2, TiO2 photocatalysis, circumneutral photo-Fenton) for the removal of organic micropollutants. *Science of The Total Environment* 2021; **761**: 143258.
8. Fernandes S, Esteves da Silva JCG, Pinto da Silva L. Life cycle assessment of the sustainability of enhancing the photodegradation activity of TiO2 with metal-doping. *Materials* 2020; **13**: 1487.

9. Li T, Zhang H, Liu Z, Ke Q, Alting L. A system boundary identification method for life cycle assessment. *The International Journal of Life Cycle Assessment* 2014; **19**: 646–60.
10. Panesar DK, Seto KE, Churchill CJ. Impact of the selection of functional unit on the life cycle assessment of green concrete. *The International Journal of Life Cycle Assessment* 2017; **22**: 1969–86.
11. Matheys J, Van Autenboer W, Timmermans J-M, Van Mierlo J, Van den Bossche P, Maggetto G. Influence of functional unit on the life cycle assessment of traction batteries. *The International Journal of Life Cycle Assessment* 2007; **12**: 191.
12. Muthu SS. 6 - Estimating the overall environmental impact of textile processing: life cycle assessment (LCA) of textile products. In: Muthu SS. Assessing the Environmental Impact of Textiles and the Clothing Supply Chain. Woodhead Publishing, Sawston UK, 2014; 105–131.
13. Speck R, Selke S, Auras R, Fitzsimmons J. Choice of Life Cycle Assessment Software Can Impact Packaging System Decisions. *Packaging Technology and Science* 2015; **28**: 579–88.
14. Speck R, Selke S, Auras R, Fitzsimmons J. Life Cycle Assessment Software: Selection Can Impact Results. *Journal of Industrial Ecology* 2016; **20**: 18–28.
15. Herrmann IT, Moltesen A. Does it matter which Life Cycle Assessment (LCA) tool you choose? – a comparative assessment of SimaPro and GaBi. *Journal of Cleaner Production* 2015; **86**: 163–9.
16. Hauschild MZ. Introduction to LCA methodology. In: Hauschild MZ, Rosenbaum RK, Olsen SI eds. *Life Cycle Assessment*. Springer, Heidelberg, Germany, 2018: 59–66.
17. Vallero DA. Chapter 5 - Waste and Biogeochemical Cycling. In: Letcher TM, Vallero DA. eds. *Waste: A Handbook for Management* (2nd edn), Academic Press, Cambridge, Massachusetts, United States, 2019: 91–125.
18. McMichael AJ, Campbell-Lendrum DH, Corvalán CF *et al. Climate Change and Human Health: Risks and Responses*. World Health Organization, 2003.
19. Zampori L, Saouter E, Schau E, Cristobal Garcia J, Castellani V, Sala S. Guide for interpreting life cycle assessment result. *Publications Office of the European Union: Luxembourg* 2016.
20. Corominas L, Byrne DM, Guest JS *et al*. The application of life cycle assessment (LCA) to wastewater treatment: A best practice guide and critical review. *Water Research* 2020; **184**: 116058.
21. Rashid SS, Liu Y-Q. Assessing environmental impacts of large centralized wastewater treatment plants with combined or separate sewer systems in dry/wet seasons by using LCA. *Environmental Science and Pollution Research* 2020; **27**: 15674–90.
22. Parra-Saldivar R, Bilal M, Iqbal HMN. Life cycle assessment in wastewater treatment technology. *Current Opinion in Environmental Science & Health* 2020; **13**: 80–4.
23. Corominas L, Foley J, Guest JS *et al*. Life cycle assessment applied to wastewater treatment: State of the art. *Water Research* 2013; **47**: 5480–92.
24. Yoshida H, Christensen TH, Scheutz C. Life cycle assessment of sewage sludge management: A review. *Waste Management & Research* 2013; **31**: 1083–101.
25. Zang Y, Li Y, Wang C, Zhang W, Xiong W. Towards more accurate life cycle assessment of biological wastewater treatment plants: a review. *Journal of Cleaner Production* 2015; **107**: 676–92.
26. Lopes TAS, Queiroz LM, Torres EA, Kiperstok A. Low complexity wastewater treatment process in developing countries: A LCA approach to evaluate environmental gains. *Science of The Total Environment* 2020; **720**: 137593.

LCA of solar photocatalytic wastewater treatment

27. Gallego-Schmid A, Tarpani RRZ. Life cycle assessment of wastewater treatment in developing countries: A review. *Water Research* 2019; **153**: 63–79.
28. Emmerson R, Morse G, Lester J, Edge D. The Life-Cycle Analysis of Small-Scale Sewage-Treatment Processes. *Water and Environment Journal* 1995; **9**: 317–25.
29. Suh Y-J, Rousseaux P. An LCA of alternative wastewater sludge treatment scenarios. *Resources, Conservation and Recycling* 2002; **35**: 191–200.
30. Dixon A, Simon M, Burkitt T. Assessing the environmental impact of two options for small-scale wastewater treatment: comparing a reedbed and an aerated biological filter using a life cycle approach. *Ecological Engineering* 2003; **20**: 297–308.
31. Renzoni R, Germain A. Life Cycle Assessment of Water: From the pumping station to the wastewater treatment plant (9 pp). *The International Journal of Life Cycle Assessment* 2007; **12**: 118–26.
32. Foley J, de Haas D, Hartley K, Lant P. Comprehensive life cycle inventories of alternative wastewater treatment systems. *Water Research* 2010; **44**: 1654–66.
33. Zhang YJ, Liu WX, Cao W Bin, Zhao CZ, Peng JJ. Study on Typical Visible Light Photocatalytic Liquid under the Life Cycle Assessment. *Materials Science Forum* 2019; **944**: 1152–7.
34. Muñoz I, Peral J, Antonio Ayllón J, Malato S, Passarinho P, Domènech X. Life cycle assessment of a coupled solar photocatalytic–biological process for wastewater treatment. *Water Research* 2006; **40**: 3533–40.
35. Chatzisymeon E, Foteinis S, Mantzavinos D, Tsoutsos T. Life cycle assessment of advanced oxidation processes for olive mill wastewater treatment. *Journal of Cleaner Production* 2013; **54**: 229–34.
36. Foteinis S, Borthwick AGL, Frontistis Z, Mantzavinos D, Chatzisymeon E. Environmental sustainability of light-driven processes for wastewater treatment applications. *Journal of Cleaner Production* 2018; **182**: 8–15.
37. Ravelli D, Dondi D, Fagnoni M, Albini A. Titanium dioxide photocatalysis: An assessment of the environmental compatibility for the case of the functionalization of heterocyclics. *Applied Catalysis B: Environmental* 2010; **99**: 442–7.
38. Muñoz I, Rieradevall J, Torrades F, Peral J, Domènech X. Environmental assessment of different solar driven advanced oxidation processes. *Solar Energy* 2005; **79**: 369–75.
39. Muñoz I, Rieradevall J, Torrades F, Peral J, Domènech X. Environmental assessment of different advanced oxidation processes applied to a bleaching Kraft mill effluent. *Chemosphere* 2006; **62**: 9–16.
40. Guinée JB, Lindeijer E. *Handbook on Life Cycle Assessment: Operational Guide to the ISO Standards.* Springer Science & Business Media, Heidelberg, Germany, 2002.
41. Hauschild MZ, Rosenbaum RK, Olsen SI. *Life Cycle Assessment.* Springer, Heidelberg, Germany, 2018.
42. Grubb GF, Bakshi BR. Life Cycle of Titanium Dioxide Nanoparticle Production. *Journal of Industrial Ecology* 2011; **15**: 81–95.
43. Jolliet O, Margni M, Charles R *et al.* IMPACT 2002+: A new life cycle impact assessment methodology. *The International Journal of Life Cycle Assessment* 2003; **8**: 324.
44. Frischknecht R, Jungbluth N, Althaus H-J *et al.* The ecoinvent Database: Overview and Methodological Framework (7 pp). *The International Journal of Life Cycle Assessment* 2005; **10**: 3–9.
45. Pennington DW, Potting J, Finnveden G *et al.* Life cycle assessment Part 2: Current impact assessment practice. *Environment International* 2004; **30**: 721–39.
46. Rebitzer G, Ekvall T, Frischknecht R *et al.* Life cycle assessment: Part 1: Framework, goal and scope definition, inventory analysis, and applications. *Environment International* 2004; **30**: 701–20.

47. Huijbregts MAJ, Steinmann ZJN, Elshout PMF *et al.* ReCiPe2016: a harmonised life cycle impact assessment method at midpoint and endpoint level. *The International Journal of Life Cycle Assessment* 2017; **22**: 138–47.
48. Bulle C, Margni M, Patouillard L *et al.* IMPACT World+: a globally regionalized life cycle impact assessment method. *The International Journal of Life Cycle Assessment* 2019; **24**: 1653–74.
49. Li C, Cui S, Nie Z, Gong X, Wang Z, Itsubo N. The LCA of portland cement production in China. *The International Journal of Life Cycle Assessment* 2015; **20**: 117–27.
50. Tan RBH, Khoo HH. An LCA study of a primary aluminum supply chain. *Journal of Cleaner Production* 2005; **13**: 607–18.
51. Burchart-Korol D. Significance of environmental life cycle assessment (LCA) method in the iron and steel industry. *Metalurgija* 2011; **50**: 205–8.
52. Ou X, Zhang X, Chang S, Guo Q. Energy consumption and GHG emissions of six biofuel pathways by LCA in (the) People's Republic of China. *Applied Energy* 2009; **86**: S197–208.
53. Liu X, Yuan Z. Life cycle environmental performance of by-product coke production in China. *Journal of Cleaner Production* 2016; **112**: 1292–301.
54. Lombardi L, Tribioli L, Cozzolino R, Bella G. Comparative environmental assessment of conventional, electric, hybrid, and fuel cell powertrains based on LCA. *The International Journal of Life Cycle Assessment* 2017; **22**: 1989–2006.
55. Sleeswijk AW, van Oers LFCM, Guinée JB, Struijs J, Huijbregts MAJ. Normalisation in product life cycle assessment: An LCA of the global and European economic systems in the year 2000. *Science of The Total Environment* 2008; **390**: 227–40.
56. Payet J. Assessing toxic impacts on aquatic ecosystems in life cycle assessment (LCA). 2004.
57. Nanaki EA, Koroneos CJ. Comparative LCA of the use of biodiesel, diesel and gasoline for transportation. *Journal of Cleaner Production* 2012; **20**: 14–9.
58. Alalm MG, Djellabi R, Meroni D, Pirola C, Bianchi CL, Boffito DC. Toward Scaling-Up Photocatalytic Process for Multiphase Environmental Applications. *Catalysts* 2021; **11**: 562.

doi: 10.2166/9781789061932_0159

Chapter 6
Analysis of patents in photocatalytic water and wastewater treatment. Part I – photocatalytic materials

Ali Zebardasti[1], Mohamad Hosein Nikfar[2], Mohammad G. Dekamin[1], Emad Sanei[3], Itzel Marquez[3] and Alireza Bazargan[4*]

[1]Pharmaceutical and Heterocyclic Compounds Research Laboratory, Department of Chemistry, Iran University of Science and Technology, Tehran 1684613114, Iran
[2]Department of Civil Engineering, K. N. Toosi University of Technology, Tehran, Iran
[3]School of Engineering and Technology, College of Science and Engineering, Central Michigan University, Mount Pleasant, Michigan 48859, USA
[4]School of Environment, College of Engineering, University of Tehran, Tehran, Iran
*Corresponding author: alireza.bazargan@ut.ac.ir

ABSTRACT
Studying the details and trends of patent publishing can help shine light on the work that inventors, institutions, and investors are carrying out in a field of technology. For this, a good set of data must first be collected and, accordingly, a picture of the future can be presented. Patent analysis methods are suitable for forecasting both the near future and a tentative trajectory of the distant future in a particular field. This chapter provides a patent analysis of photocatalytic materials for water and wastewater treatment. Starting with an overview of trend analysis in patents, the results of the analysis of patent registration over time and the activity of companies, countries, and top researchers in this field are analyzed. The results show that Japan was a pioneer in the area in the 90s and early 2000s. Since then, the Chinese have dominated the field. In fact, currently, more patents on photocatalytic materials for water and wastewater treatment are filed in China than in the rest of the world combined.

6.1 INTRODUCTION
Patent registration documents contain valuable technical and legal information that allow their analysis. Before analyzing the patents in photocatalytic materials for water and wastewater treatment, a description of patent registration documents and their classification is presented. The main features of a patent registration are:

- *The Extent of Information*
  Millions of patents have been registered across the globe, providing important technical details. Although most patents are only several

© 2022 The Editors. This is an Open Access book chapter distributed under the terms of the Creative Commons Attribution Licence (CC BY-NC-ND 4.0), which permits copying and redistribution for noncommercial purposes with no derivatives, provided the original work is properly cited (https://creativecommons.org/licenses/by-nc-nd/4.0/). This does not affect the rights licensed or assigned from any third party in this book. The chapter is from the book Photocatalytic Water and Wastewater Treatment, Alireza Barzagan (Ed.).

pages, some patents may extend for hundreds of pages, and in some rare cases, the patent and its supporting data can reach thousands of pages.

- *The Sole Source*
  As inventions are commercially sensitive, patent registration documents are some of the only information sources for modern innovations in most cases. Research shows that as much as 80% of technical disclosures revealed in patents cannot be found in other sources, such as peer-reviewed journal articles [1].

- *Detailed Description*
  The content of each document ought to include a detailed description of the invention. Those descriptions must have sufficient clarity to enable any subject specialist of that field to reproduce the invention with minor trial and error. Although this is theoretically true, in reality, the information provided in patents is often cloaked in some levels of ambiguity, and details are sometimes intentionally kept vague. Patent applicants often want to keep their claims as broad as possible, which can be achieved by choosing words with a convenient degree of semantic indeterminacy [2]. Such vagueness has caused some serious issues and has even fueled anti-patent opinions [3].

- *Unified Writing Structure*
  The unified standard writing structure of patent documentation facilitates its understanding and extraction of valuable information. Other forms of scientific literature, such as peer-reviewed journals, follow other writing structures that are not the same as those of patents. Unlike patents, reports and internal documents often do not have a unified structure and can change based on the authors' tastes.

- *Accessibility*
  One of the most important features in patent registrations is accessibility to the complete content via the internet. Nowadays, there are various databases that house millions of patents. For instance, Google patents currently boasts an index of over 120 million patents from more than 100 patent offices from around the world [4]. The website of the European Patent Office, Espacenet, also holds more than 130 million patents (as of August 2021) [5].

- *Standard Classification*
  Standard classification of patent registration documents is based on the Strasbourg agreement – a multilateral treaty agreed upon in 1971 – and the World Intellectual Property Organization (WIPO), acknowledged by more than 100 countries. The patent registration offices of the bulk of nations, as well as the WIPO, comply with international patent classifications in accordance with the Strasbourg agreement.

According to the 7th edition of the International Patent Classification (IPC), in 2000, the standard classification had 8 sections, 120 classes, 628 subclasses, and 69,000 groups. Such extensive classification provides quick access to the required information. A complete classification comprises the combined symbols representing the section, class, subclass, and main group or subgroup.

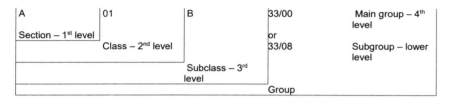

Figure 6.1 Example of standard patent classification Reprinted from [7]. A: Section 'A' human necessities. A01: agriculture; forestry; animal husbandry; hunting; trapping; fishing. A01B: soil working in agriculture or forestry; parts, details, or accessories of agricultural machines or implements, in general. A01B33/00: Tilling implements with rotary-driven tools. A01B33/08 (a subgroup or lower level of the main group): Tools and details, such as adaptations of transmissions or gearings.

An example of the classification can be seen in Figure 6.1 for the IPC class A01B33/00 or A01B33/08, each containing thousands of patents. Nonetheless, some patents may fit under different classes, and in a thorough search, more than one class needs to be considered for a particular technology [6].

6.2 PATENT REGISTRATION DOCUMENT ANALYSIS

Databases are an excellent source for analyzing patents and their registration documents. Policymakers, private companies, researchers, and academics are target audiences of patent analysis information. Some of the valuable outcomes that patent analysis provides can be listed as:

- Avoiding repetitive and redundant research studies, thus reducing research costs
- Initiating future research from a higher level
- Implementing novel methodologies
- Protecting intellectual property rights
- Staying up to date with the latest achievements and identifying lucrative investments

Patent registration organizations store information such as expiration and publication date, inventors' personal information, and international classification number of the patents. Some of this information can be extracted and analyzed to pave the way for future development. The cumulative number of compiled patents reflects the technology lifecycle, from the expense of the research and development phase to the decline phase.

The patenting process starts with the initial idea but requires additional research and understanding of how patents are used and registered. Once an understanding of the patenting process is gained, additional research and investigations will be required to prepare a suitable patent to be submitted to the relevant office. Meanwhile, funding for the project needs to continue, both for additional research as well as for the patenting and legal fees, which may very well run into thousands of dollars per patent [8].

162   Photocatalytic Water and Wastewater Treatment

Information extracted from other patents is analyzable and usable when applying for patent registration. For example, learning about new innovations helps foster new ideas and can help understand the trend that a particular technology is taking to focus on patenting ideas or products that will be of importance in the future. Patent analysis can also show the orientation of active companies in each field of technology and help select appropriate partners to develop innovations. The geographical spectra for inventions in a particular technological area can also help understand regional trends (e.g., US patents versus Chinese patents).

It should be noted that, in countries such as the United States, provisional and non-provisional patents have different registration and publication processes. Non-provisional patent applications (generally referred to as patents) are published after 18 months from their earliest submission for inspection. Whenever an inventor files a non-provisional patent application, the US Patent Office will publish it for public viewing even if it is not accepted as an issued patent. On the other hand, provisional patent applications are never published and cannot be found online unless converted into non-provisional patents down the line. If the status of the provisional patent is changed to non-provisional, the 18 months is calculated from the date of the first application.

Provisional patent applications are cheaper and less critical because they are not examined by a US Patent Office Patent Examiner. They remain in a 'patent pending' status for 12 months unless they are converted into a formal non-provisional patent application by the inventor. The provisional patent is intended to provide a year for the inventor to find potential investors, test commercial feasibility, consult with licensing bodies, determine a marketing and sales strategy, and conduct additional experiments before committing to the expensive formal non-provisional patent application [9].

In 2019, a total of 3.2 million patents were filed worldwide. Under 70% were filed by residents of countries in which they were filed, and over 30% were filed by non-residents. In fact, more than 50% of the patents filed in the US and Europe are filed by non-residents. Figure 6.2 shows the top 20 patent application offices in the world by number of applications and their relative resident and non-resident shares. Note that countries such as Australia, Hong Kong, and Mexico receive more than 90% of their applications from non-residents.

6.2.1 The main steps of patent analysis

The first step for analyzing patents in a particular field, such as photocatalytic water and wastewater treatment, is to extract the most relevant registered patents and use these documents to extract information, which is then interpreted and analyzed. The main steps of this process can be generally defined as:

- Defining specific keywords for an initial search
- Using the keyword search to extract the relevant classification codes
- Finding and choosing the relevant patents to be included in the analysis
- Text mining for extracting relevant information from the patents

The first and most crucial part of searching for relevant patents is to identify specific keywords and their combinations in a given field. In this regard, relevant keywords can be extracted by studying the most significant available resources

Analysis of patents in photocatalytic water 163

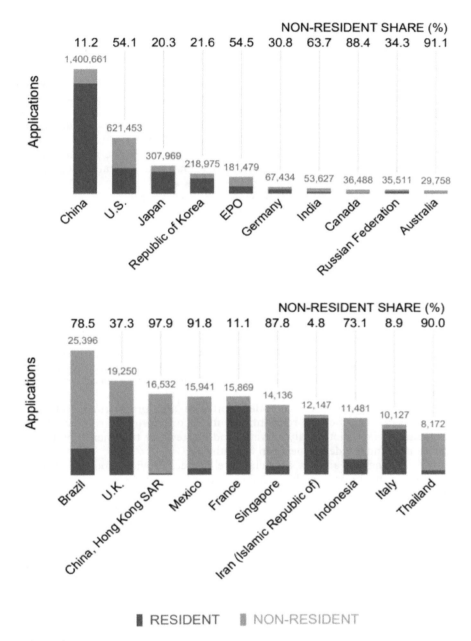

Figure 6.2 Top 20 patent application offices in the world by share of resident and non-resident filings. EPO is the European Patent Office. In general, national offices of the EPO member states receive lower volumes of applications because applicants may apply via the EPO to seek protection within all EPO member states Reprinted from [10].

and conducting detailed research, including patent documents, articles, books, news, and technical reports for later use in the following stages of the patent analysis process. When looking for something very specific, this task becomes even more challenging.

After the keywords are used for an initial search, the relevant codes in which the required patent may exist start to emerge. Nonetheless, keyword searches may lead to irrelevant or unsuitable results because regional differences and personal preferences can result in the use of different terms when describing the same concepts. In chemistry, for example, a chemical mixture can have numerous different names. Finding the relevant codes for the inventions that one is looking for is an iterative process that may take several searching and tuning steps.

After an initial number of keyword searches have been conducted, the relevant codes can be distinguished and used for further in-depth searches. For photocatalytic materials in water and wastewater treatment, the top IPC and Cooperative Patent Classification (CPC) codes that were found are listed in Tables 6.1 and 6.2, respectively. The CPC is an extension of the IPC and is jointly managed by the EPO and the US Patent and Trademark Office. CPCs may be more up-to-date, specific, and detailed than IPCs; however, IPCs are used in a lot more countries and are a lot more recognized internationally. Interestingly, only a few rows in Tables 6.1 and 6.2 include the word 'photocatalyst' or 'photocatalytic' within them. The majority of rows do not exhibit any apparent connections to the topic. It was only through a keyword search and screening of the results that these relevant codes were found. As previously shown in Figure 6.1, the codes for the main groups end in '00' while the codes for the subgroups end in other numbers, each placed under the associated main group within the hierarchy.

After finding the relevant codes, software and programing techniques can be used to extract the relevant patents according to their keywords and applications. For photocatalytic water and wastewater treatment, the patents were found and divided to perform text mining.

Text mining, an area in artificial intelligence, provides tools to discover information by automatically extracting data from various written sources. It includes all the activities required to gain information from the text. Textual data analysis by machine learning techniques, intelligent retrieval of intelligent data, natural learning processing, and other relevant methods are within the scope of text mining methods.

Examples of text mining applications include classifying documents into a set of specific topics (supervised learning), categorizing documents so that each category has the same meaning (clustering and unsupervised learning), and finding documents meeting the needs of searching criteria (information retrieval). However, before using text mining methods, it is best to structure primarily unstructured or semi-structured texts. At this stage, data mining at different timespans is conducted in processing statistical texts, discovering data through the texts, automatic analysis of text, and processing natural language.

Previous literature has thoroughly assessed various text mining methods for patent analysis. For instance, the rationale behind vector space models,

Analysis of patents in photocatalytic water 165

Table 6.1 The most frequent IPC patent codes used for patents relevant to photocatalytic water and wastewater treatment.

	Code Used	% of Patents	Description
Main groups	C02F1/00	25	Treatment of water
	B01J35/00	11	Catalysts
	C02F101/00	9	Nature of the contaminant
	C02F9/00	8	Multistep treatment of water
	B01J23/00	7	Catalysts comprising metals or metal oxides or hydroxides
	B01J27/00	7	Catalysts comprising the elements or compounds of halogens
	B01J21/00	6	Catalysts comprising the elements
	B01J37/00	5	Processes
	B01D53/00	4	Separation of gases or vapors
	B01J31/00	3	Catalysts comprising hydrides
	C01B3/00	2	Hydrogen
	C02F103/00	2	Nature of water
	C01G23/00	2	Compounds of titanium
	A61L9/00	2	Disinfection
	C09D5/00	2	Coating compositions
	B01J20/00	2	Solid sorbent compositions or filter aids
Main subgroups	C02F1/30	16	By irradiation
	C02F1/32	14	With ultra-violet light
	C02F1/72	10	By oxidation
	B01J35/02	9	Solids
	B01J21/06	6	Silicon
	C02F101/30	6	Organic compounds
	C02F9/08	5	At least one step being a physical treatment
	B01D53/86	5	Catalytic processes
	C02F101/38	3	Containing nitrogen
	B01J27/24	3	Nitrogen compounds
	C02F1/28	3	By sorption
	C02F9/14	3	At least one step being a biological treatment
	C02F101/34	3	Containing oxygen
	C01B3/04	3	By decomposition of inorganic compounds
	B01J35/10	2	Characterized by their surface properties

The '% of patents' shows the percentage of patents from the final selected list of patents that fit under a particular code.

166 **Photocatalytic Water and Wastewater Treatment**

Table 6.2 The most frequent CPC patent codes used for patents relevant to photocatalytic water and wastewater treatment.

	Code Used	% of Patents	Description
Main groups	Y02W10/00	17	Technologies for wastewater treatment
	C02F1/00	12	Treatment of water
	B01J35/00	10	Catalysts
	C02F2305/00	9	Use of specific compounds during water treatment
	Y02E60/00	7	Enabling technologies or technologies with potential or indirect contribution to GHG emissions mitigation
	B01J37/00	6	Processes
	B01J21/00	5	Catalysts comprising the elements
	C02F2101/00	4	Nature of the contaminant
	B01J23/00	4	Catalysts comprising metals or metal oxides or hydroxides
	C02F2201/00	4	Apparatus for treatment of water
	B01D53/00	3	Separation of gases or vapors. Recovering vapors of volatile solvents from gases. Chemical or biological purification of waste gases.
	C02F2103/00	3	Nature of the water
	B82Y30/00	2	Nanotechnology for materials or surface science
Main subgroups	Y02W10/37	19	Using solar energy
	B01J35/004	11	Photocatalysts
	C02F2305/10	11	Photocatalysts
	C02F1/725	9	By catalytic oxidation
	C02F1/32	7	With ultra-violet light
	Y02E60/36	7	Hydrogen production from non-carbon containing sources
	B01J21/063	6	Titanium; oxides or hydroxides thereof
	C02F1/325	5	Irradiation devices or lamp constructions
	C02F1/30	3	By irradiation
	B01J35/002	3	Catalysts characterized by their physical properties
	B82Y30/00	3	Nanotechnology for materials or surface science
	Y02W10/10	3	Biological treatment of water
	B01D2255/802	3	Photocatalytic
	B01J35/0013	2	Colloids

The '% of patents' shows the percentage of patents from the final selected list of patents that fit under a particular code.

latent semantic analysis, and probabilistic topic models has been explained. It has been shown that choices in terms of algorithms, pre-processing, and calculation options have significant consequences in text mining outcomes. Therefore, the technicalities and details of the text mining step are essential in determining the outcomes [11]. Since a large amount of human judgment is required for screening early ideas and patents, which may prove important in the future, text mining techniques have been proposed for screening purposes. For instance, in one study, keyword vectors were constructed from patents [12]. In this case, the k-nearest neighbors algorithm was used to capture the relationships between the keyword vectors and the numbers of forwarding citations of the patents.

Another method is natural language processing. During natural language processing, the process of word sense disambiguation (WSD) is designed to determine which grammatical meaning of a given word is being invoked in a given context. Since the number of patent documents has grown tremendously due to rapid technological advances, companies are struggling with how to use the large numbers of patent documents to find new business opportunities and avoid conflicts with existing patents. Natural language processing can help solve such problems [13].

Text mining is concerned with seeking out useful information from unstructured textual data, in this case, from patents pertaining to photocatalytic water and wastewater treatment. The text mining process can help perform fast and accurate analyses of patents' contents covering the whole context and providing a detailed summary. In this study, text mining was used to review essential points on patents associated with photocatalytic materials in water treatment.

6.3 PATENT ANALYSIS FOR PHOTOCATALYTIC MATERIALS IN WATER AND WASTEWATER TREATMENT

This section is dedicated to a comprehensive search regarding photocatalytic materials in water and wastewater treatment. The search and data extraction were performed in English; therefore, only patents with identifying information in English (such as title and abstract) were considered.

After searching through databases for subject-related patents, a total of 11,527 patents relevant to photocatalytic materials for water and wastewater treatment were extracted. Of these patents, 52.18% were not granted, while 47.82% were successful. This means that a total of 5512 granted patents were found. However, it is important to note that not all patents have industrial applications. In fact, most patents filed never make it off the shelf into practical applications. A strength weakness opportunity threat (SWOT) analysis of photocatalytic processes for environmental remediation has recently identified the relative advantages and disadvantages impacting the implementation of photocatalytic technologies [14], confirming that many photocatalytic technologies are not applied, even if they are patented. Furthermore, since the theoretical photocatalytic activity achieved in laboratories can never be achieved in industrial applications, researchers have even defined an upper practical limit and its calculation method [15].

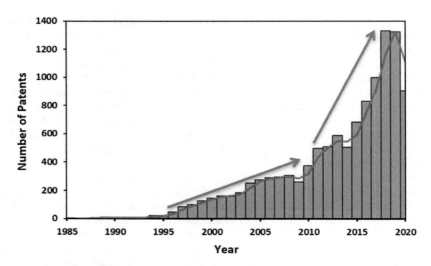

Figure 6.3 Number of filed patents per year in the field of photocatalytic materials for water and wastewater treatment.

As shown in Figure 6.3, the trend for photocatalytic material patenting started as early as 1985; however, until the mid-90s, the number of patents in this field was minimal. From 1995 to 2008, a period of growth can be seen with an increase in slope in the number of patents registered each year. Nevertheless, after 2008 the trend presents a massive growth, indicating that the interest in this technical area is increasing. There are many reasons behind attracting more researchers and inventors to this field, such as increased awareness of water quality and water pollution, ever-growing water scarcity around the globe, and more stringent effluent standards. The drop in 2020 should not be alarming, since the analysis was carried out in the middle of that year.

Figure 6.4 shows the total number of patents relevant to photocatalytic materials in water treatment filed by inventors from a particular country as of mid-2020. China is at the top of the list with 4811 patents, followed by Japanese inventors with 1393 patents. Inventors from South Korea and the USA with 499 and 293 patents, respectively, are third and fourth. Of the top five countries in the list, four are east-Asian, showing the dominance of inventors from this region. In contrast, Figure 6.5 shows the location where the patents have been registered as of mid-2020 (i.e., the patent destination). In other words, Figure 6.5 shows how many patents have been filed in the patent office of a particular country or territory. As per the figure, the largest number of patents were filed in China (8159 patents), followed by Japan with 1553 patents. This means that China is the top destination for patent registration, presumably due to its enormous target market. In fact, the surge in Chinese patents in various fields is a widespread phenomenon [16, 1] (sometimes referred to as one of the aspects of Chinese technological catch-up on the world stage) and has been reported and investigated thoroughly [18–26].

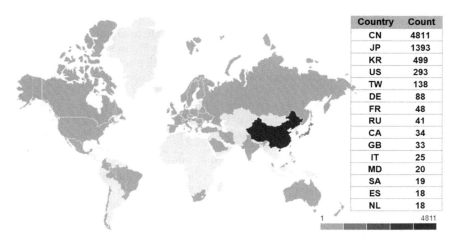

Figure 6.4 Total number of filed patents in the field of photocatalytic materials for water and wastewater treatment by inventors from each country as of mid-2020. CN: China; JP: Japan; KR: South Korea; US: United States; TW: Chinese Taipei (Taiwan); DE: Germany; FR: France; RU: Russia; CA: Canada; GB: Great Britain; IT: Italy; MD: Moldova; SA: Saudi Arabia; ES: Spain; NL: the Netherlands.

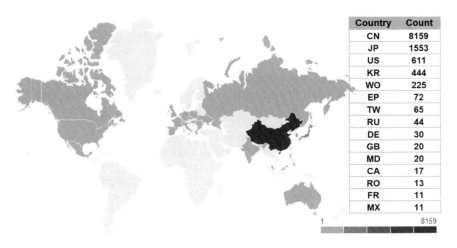

Figure 6.5 Total number of patents filed in the field of photocatalytic materials for water and wastewater treatment in each country or territory as of mid-2020. CN: China; JP: Japan; US: United States; KR: South Korea; WO: Patent Cooperation Treaty (PCT) also known as International Application; EP: European Patent Office; TW: Chinese Taipei (Taiwan); RU: Russia; DE: Germany; GB: Great Britain; MD: Moldova; CA: Canada; RO: Romania; FR: France; MX: Mexico.

A total of 4811 patents in the field have been filed by Chinese inventors worldwide (as shown in Figure 6.4). However, it is safe to assume that not all of these were filed in their home country. By comparing Figures 6.4 and 6.5, it can be estimated that at least half of the patents regarding photocatalytic materials for water and wastewater treatment that have been filed in China belong to foreign nationals.

The United States with 611 patents is the third most popular destination, followed by South Korea with 444 patents. Therefore, although we had seen that South Korean inventors are more active in the field, the US is prioritized over South Korea as a patent destination.

As explained earlier, the European Patent Office receives more applications than the national offices of member states because applicants may apply via the EPO to seek protection within all EPO member states. While there is no such thing as a worldwide patent, there is what might be compared to a global patent application which, if successful, can lay the grounds for the application to be processed in many countries around the world. A patent filed like this is called a World Intellectual Property Organization patent application or a Patent Cooperation Treaty (PCT) application, which is the name of the international treaty that authorizes the filing of this single application. In other words, it is possible to file a single worldwide patent application, but it is *not* possible to obtain a single worldwide patent. This is because patents are granted by individual countries, not by any international authority. The purpose of the PCT application is to streamline as many overlapping procedures as possible. Since individual countries have different patent laws, a uniform worldwide patent is impossible. For example, genetically bioengineered microorganisms are not patentable in many countries, while it is possible to patent them in the United States.

As mentioned, there are two main steps to the PCT procedure. The first step involves filing an international application. The second step occurs if the applicant chooses to obtain an enforceable patent in any of the signatory countries, where it is evaluated under the laws of that particular country. Thus, there is an international phase and a national phase to the PCT process [27]. Applications through the PCT are designated with a WO prefix in Figure 6.5 and are standing in fifth place.

Figure 6.6 demonstrates the trend of patent registration in target countries over time. As evident, throughout the 90s and early 2000s, the Japanese market was the target of most photocatalytic material patents. It is as if during the time that the Japanese were exploring various photocatalytic materials, the rest of the world was lagging. However, China started to emerge in the early years of the millennium. From the second half of the 2000s onwards, China began to dominate in the field of photocatalytic material patents for water and wastewater treatment. When the activity of inventors was also investigated (not shown in the figure), a similar trend was observed: the waning of the Japanese and the increase of the Chinese in the field.

6.3.1 The registrants of patents in the field of photocatalytic materials for water and wastewater treatment

One of the crucial areas in patent analysis is identifying the major patent registrants. It is possible to extract this information from the database of

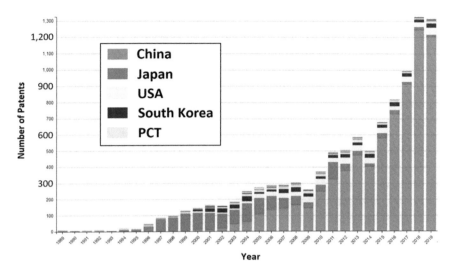

Figure 6.6 Patents filed in the field of photocatalytic materials for water and wastewater treatment in each destination country. Green: China; Red: Japan; Yellow: USA; Dark Blue: South Korea; Light Blue: World Intellectual Property Organization (PCT).

relevant patents created. Knowing who has registered the patents can be useful in various ways, for example, for identifying potential threats, technology acquisition, research and development cooperation, and other goals.

The results of the patent analysis in the field showed that companies registered 6013 patents, while 4626 patents were registered by research and academic institutions. The balance between both, private companies and research institutes, shows that the field is more or less balanced. Table 6.3 shows the list of top research institutes and universities active in this field. Jiangsu University in China emerged as the leading registrant. Many other Chinese universities appear on the list of top owners.

The analysis also showed that, out of the total number of patents registered by Chinese registrants, 3845 patents belong to universities, while Chinese companies have registered only 906 patents. This means that Chinese universities are more active in this regard compared to Chinese companies. The opposite is true in other countries where companies lead the charge. For instance, out of 1393 patents registered by Japan, only 88 patents have been registered in academies and institutes, and the rest belong to private companies. Thus, unlike China, in Japan, the activity of companies in this field is greater than that of academic and research centers.

Table 6.4 shows the top companies active in this field worldwide. According to the results, the Japanese company TOTO holds the greatest number of patents among companies in photocatalytic materials for water and wastewater treatment. Matsushita Electric, now Panasonic, a Japanese superpower, is the second-largest patent holder. The third-largest patent owner is the Chinese company Chengdu New Keli Chem Sci Co., which, according to Figure 6.7,

Photocatalytic Water and Wastewater Treatment

Table 6.3 List of research institutes and universities that have registered the greatest number of patents in the field of photocatalytic materials for water and wastewater treatment (abbreviations taken directly from the analysis).

Academic Registrants of Patents	Count
UNIV JIANGSU	124
UNIV NINGBO	103
UNIV CHANGZHOU	89
UNIV HOHAI	80
UNIV NANJING	75
UNIV ZHEJIANG	71
UNIV SHANGHAI	60
UNIV SHANGHAI JIAOTONG	60
UNIV TONGJI	57
UNIV TIANJIN	56
UNIV FUZHOU	56
UNIV SHAANXI SCIENCE & TECH	54
UNIV SHANDONG	50
UNIV HENAN NORMAL	50
UNIV SOUTH CHINA TECH	49
UNIV WUHAN TECH	44

does not have a history in this field and suddenly registered all of its patents in 2018. Meanwhile, the Japanese companies TOTO, Matsushita Electric, and Sharp K.K. have been working in this field for decades.

Evidently, the strongest annual showing belongs to TOTO in 2011; they filed more patents in the field in this one year alone than any other company has over their entire lifespan. Furthermore, although China surpasses Japan's number of patents, Japan has been working on these technologies for a longer period. In addition, Japan's technology comes mainly from private companies, increasing the chance of industrial and semi-industrial scale applications.

The destination country for filing patents by the top owners in photocatalytic materials for water and wastewater treatment is shown in Figure 6.8. As the results indicate, the most favorable target destination for these corporations is Japan, followed by China and the United States.

In some cases, patent owners cooperate to purchase or obtain patents, which indicates that these owners collaborated in research and development or in other areas. In general, collaborations are limited because each patent holder wishes to maximize the exclusivity of their technology, that is, wishes to monopolize it. Through monopolization or trade secrets, companies seek to preserve their competitive advantage and prevent knowledge leakage to their competitors. Figure 6.9 shows the cooperation in patent registration between companies in the field. According to this figure, although cooperation exists between companies, it is limited.

Analysis of patents in photocatalytic water

Table 6.4 List of companies that have registered the greatest number of patents in the field of photocatalytic materials for water and wastewater treatment (abbreviations taken directly from the analysis).

Company Registrants of Patents	Count
TOTO LTD	163
MATSUSHITA ELECTRIC IND CO LTD	57
CHENGDU NEW KELI CHEM SCI CO	40
SHARP KK	39
TOSHIBA KK	28
JAPAN SCIENCE & TECH AGENCY	25
AGENCY IND SCIENCE TECHN	25
UBE INDUSTRIES	25
EBARA CORP	21
DAINIPPON PRINTING CO LTD	21
TORAY INDUSTRIES	19
BRIDGESTONE CORP	19
KANAGAWA KAGAKU GIJUTSU AKAD	19
MEIDENSHA ELECTRIC MFG CO LTD	19
WUHAN DONGCHUAN WATER ENV LTD CO	19
CHINESE ACAD INST CHEMISTRY	18
NAT INST OF ADV IND & TECHNOL	18
PROCTER & GAMBLE	17
MITSUBISHI HEAVY IND LTD	17

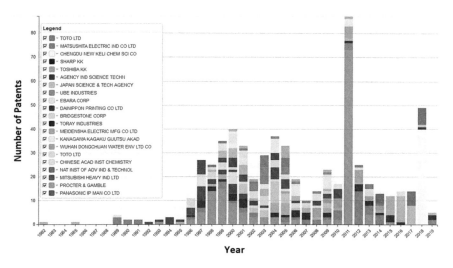

Figure 6.7 Most active companies in filing patents in photocatalytic materials for water and wastewater treatment by year.

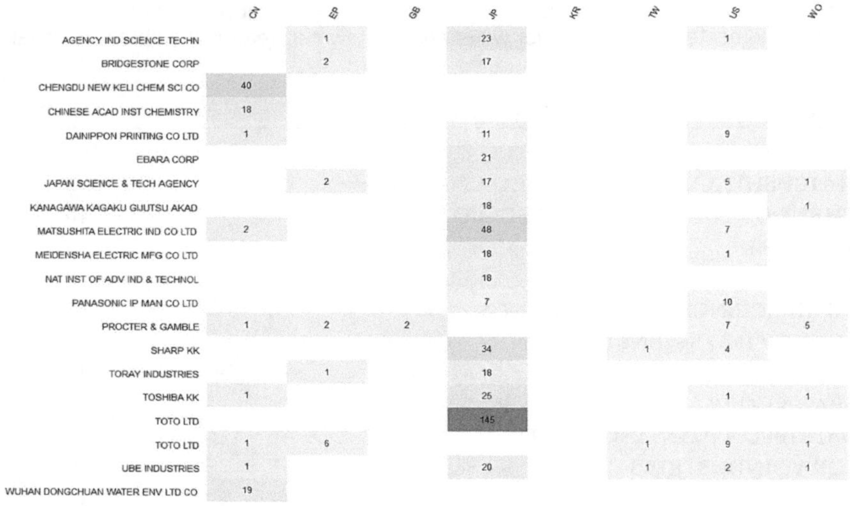

	CN	EP	GB	JP	KR	TW	US	WO
AGENCY IND SCIENCE TECHN		1		23			1	
BRIDGESTONE CORP		2		17				
CHENGDU NEW KELI CHEM SCI CO	40							
CHINESE ACAD INST CHEMISTRY	18							
DAINIPPON PRINTING CO LTD	1			11			9	
EBARA CORP				21				
JAPAN SCIENCE & TECH AGENCY		2		17			5	1
KANAGAWA KAGAKU GUUTSU AKAD				18				1
MATSUSHITA ELECTRIC IND CO LTD	2			48			7	
MEIDENSHA ELECTRIC MFG CO LTD				18			1	
NAT INST OF ADV IND & TECHNOL				18				
PANASONIC IP MAN CO LTD				7			10	
PROCTER & GAMBLE	1	2	2				7	5
SHARP KK				34	1		4	
TORAY INDUSTRIES		1		18				
TOSHIBA KK	1			25			1	1
TOTO LTD				145				
TOTO LTD	1	6				1	9	1
UBE INDUSTRIES	1			20		1	2	1
WUHAN DONGCHUAN WATER ENV LTD CO	19							

Figure 6.8 The destination country for filing patents by the top patent holders in photocatalytic materials for water and wastewater treatment. CN: China; EP: European Patent Office; GB: Great Britain; JP: Japan; KR: South Korea; TW: Chinese Taipei (Taiwan); US: United States; WO: Patent Cooperation Treaty (PCT) also known as International Application.

Another important analysis for patents is the name of the top researchers and scholars in that field, designated as the 'inventors' of the technology. A researcher owning numerous patents in a particular technological field can play a critical role in its implementation. Figure 6.10 shows the top inventors in photocatalytic materials for water and wastewater treatment. The leading researchers on the list include Li Rongsheng, Ren Yuanlong, Ning Gan, and Wang Dongjie, researchers at Ningbo University, as well as Hayakawa Makoto, Kameshima Junji, and Takaki Yoji from TOTO LTD's R&D team. The top 11 names in Figure 6.10 have been named as inventors a total of 1000 times altogether.

6.3.2 Key extracted concepts

One of the most significant outputs of patent analysis is extracting critical concepts in a particular field. Information such as technology sub-disciplines, materials used, and applications of the inventions can be extracted from the text mining process. Table 6.5 shows the top 50 most common words or phrases used in patents in photocatalytic materials for water and wastewater treatment. According to these results, titanium dioxide is by far the most widely used word in the field. Polymers such as polyethylene glycol and polyvinyl alcohol are also seen in the list. Furthermore, the emergence of words relevant to nanotechnology

Analysis of patents in photocatalytic water

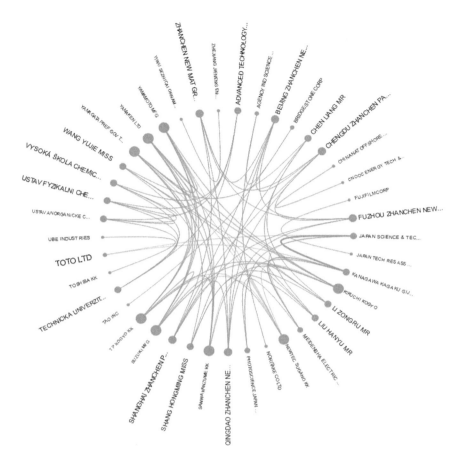

Figure 6.9 Cooperation between companies in filing patents in photocatalytic materials for water and wastewater treatment.

shows the close relationship of nanotechnology with photocatalysis, which will be discussed in depth in the following chapter.

Table 6.6 shows the results of intelligent text mining on words identified as problems to be solved discussed in patents in this area. The problems listed are in order of the number of mentions in the patents.

Another analysis showed the main subject areas relevant to the application of the patents. Figure 6.11 shows that the most common application perceived for the patents is environmental technology with 6491 patents, followed by chemical engineering with 5936 patents. Various other industries, such as materials and medicine, emerge also as technology applications.

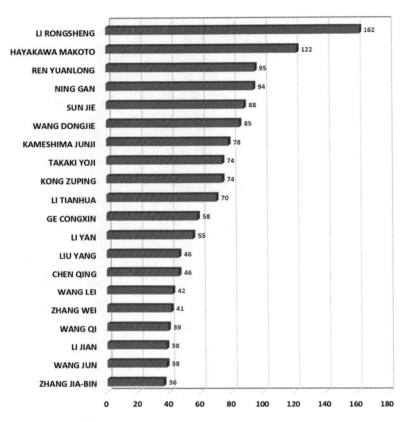

Figure 6.10 The top inventors of patents for photocatalytic materials for water and wastewater treatment.

6.4 CONCLUSION

In this chapter, patent analysis regarding photocatalytic materials for water and wastewater treatment was carried out. First, a detailed search within the literature showed the frameworks of this technology and the most relevant keywords. After the keywords were identified, the relevant international classification codes (both the IPC and CPC) in this field of technology were extracted. A total of 11,527 patents were extracted by searching databases which included both the granted and non-granted patents. The main results of the patent analysis are as follows:

- Patents regarding the photocatalytic treatment of water and wastewater date back to the 80 s. Up until the mid-90 s, activity in the field was very limited. From 1995 to 2008, activity started to increase at a steady rate. From 2008 until today, there has been a sharp increase in the filing of patents in this field.

Table 6.5 Results of patent text mining in terms of frequently repeated words.

Rank	Word	Number of Patents Using This Word or Phrase	Rank	Word	Number of Patents Using This Word or Phrase
1	Titanium dioxide	1822	26	Chemical reaction	189
2	Carbon dioxide	588	27	Catalyst particles	186
3	Energy consumption	531	28	Nano TiO_2	181
4	Hydrogen peroxide	518	29	Titanium dioxide particles	177
5	Ultraviolet lamp	513	30	Polyethylene glycol	174
6	Ethyl alcohol	482	31	TiO2 particles	173
7	Ultraviolet rays	376	32	Titanium dioxide powder	170
8	Valence band	359	33	Reaction vessel	152
9	Hydroxyl radicals	349	34	Silicon dioxide	150
10	Reaction kettle	336	35	Adsorption capacity	149
11	Sodium hydroxide	334	36	Butyl titanate	144
12	Mercury lamp	331	37	Carbon fiber	141
13	Titanium dioxide photocatalyst	324	38	Silver nitrate	140
14	TiO_2 photocatalyst	252	39	Nanometer TiO_2	138
15	Nanometer titanium dioxide	252	40	Nano titanium dioxide	136
16	Tetrabutyl titanate	250	41	Delivery port	134
17	Methyl orange	233	42	Absorption spectrum	133
18	Photocatalyst particles	225	43	Carbon adsorption	133
19	Methylene blue	218	44	Redox reaction	129
20	Titanium tetrachloride	216	45	TiO_2 powder	120
21	Quartz glass	216	46	Silica gel	118
22	Ethylene glycol	210	47	Nanometer photocatalyst	118
23	Sol gel	200	48	Xenon lamp	117
24	Polyvinyl alcohol	191	49	Titanium alkoxide	117
25	Reaction tank	190	50	TiO_2 photocatalysis	116

Table 6.6 The results of text mining in terms of recurring problems mentioned in patents.

Rank	Word	Number of Patents Using This Word or Phrase	Rank	Word	Number of Patents Using This Word or Phrase
1	Pollution problem	84	26	Mass transfer problem	9
2	To recycle	45	27	Treatment problem	9
3	To degrade	41	28	To treat	8
4	To distinguish	37	29	Corrosion problem	8
5	To control	23	30	Health problem	8
6	To process	22	31	To biodegrade	8
7	To remove	20	32	Oxidation technology	8
8	To separate	18	33	To disperse	7
9	Water pollution problem	16	34	Odor problem	7
10	To dissolve	16	35	To generate sufficient acid	7
11	To achieve	16	36	Series problem	7
12	To decompose	15	37	To carry	6
13	Energy problem	13	38	To degrade the organic wastewater	6
14	Quantum efficiency problem	13	39	Energy utilization problem	6
15	To degrade the organic matter	13	40	Agglomeration problem	6
16	To degrade organic pollutants	13	41	To handle	6
17	Selecting the model	13	42	To degrade the organic wastewater method	6
18	Degradation reaction endpoint	13	43	Degradation reaction endpoint time	6
19	To degrade the organic matter	12	44	Protection problem	5
20	To degrade the organic waste water	12	45	To separate the recovery	5
21	Safety problem	12	46	To obtain	5
22	To use	11	47	Photocatalyst	5
23	Separation problem	11	48	To clean	5
24	Recovery problem	10	49	Stability problem	5
25	To recover	10	50	Mass transfer problem	9

Figure 6.11 The most relevant subject areas to the application of the patents in photocatalytic materials for water and wastewater treatment.

- The top countries for patent registration in photocatalytic materials in water treatment are China, Japan, and South Korea. China shows the largest number of patents, and its activity has increased with time.
- Patent registration was analyzed in terms of destination countries, indicating that the top destinations for patent registration are China, Japan, and the United States, followed by South Korea. China's large number of patents suggests a large and attractive market.
- Several Chinese universities are the top owners of patents in this field. Jiangsu University from China emerged as the leading owner in this group. The top companies in this area are the Japanese companies TOTO and Matsushita Electric (now Panasonic), and the Chinese company Chengdu New Keli Chem Sci Co.

180 Photocatalytic Water and Wastewater Treatment

- The results of intelligent text mining demonstrate that the titanium dioxide photocatalyst has emerged as the most widely mentioned photocatalyst in the patents. Polymers such as polyethylene glycol and polyvinyl alcohol are also used. Additionally, words regarding nanotechnology as well as natural sunlight frequently appear, which is the subject of the next chapter.
- The most relevant applications for the patents analyzed were environmental technology and chemical engineering. The most commonly occurring problems attempted to be solved with these technologies are environmental issues such as pollution and waste.

REFERENCES

1. Asche G. "80% of technical information found only in patents" – Is there proof of this [1]? *World Patent Information* 2017; **48**: 16–28.
2. Arinas I. How Vague Can Your Patent Be? Vagueness Strategies in U.S. Patents. *HERMES - Journal of Language and Communication in Business* 2012; **48**: 55–74.
3. Gollin M. The cost of vague patents. *Nature* 2008; **454**: 164–5.
4. Google Help. About Google Patents - Coverage. https://support.google.com/faqs/answer/7049585?hl=en (accessed August 2, 2021).
5. European Patent Office. Espacenet patent search. https://www.epo.org/searching-for-patents/technical/espacenet.html (accessed August 2, 2021).
6. White M. Patent Searching: Back to the Future How to Use Patent Classification Search Tools to Create Better Searches. *Proceedings of the 1st Canadian Engineering Education Association (CEEA) Conference.* Queens University Kingston, 2010.
7. World International Property Organization. *Guide to the International Patent Classification.* 2020.
8. Peichel J, Bazargan A. Research and Development Management. In: Bazargan A, ed. *A Multidisciplinary Introduction to Desalination.* River Publishers, Delft, The Netherlands, 2018: 295–321 .
9. Kaufhold & Dix. Are Provisional Patent Applications Published? 2020. https://www.kaufholdpatentgroup.com/patent-search-are-provisional-patents-published/ (accessed August 4, 2021).
10. World International Property Organization. *World Intellectual Property Indicators 2020.* 2020.
11. Van Looy B, Magerman T. Using Text Mining Algorithms for Patent Documents and Publications. In: Glänzel W, Moed HF, Schmoch U, Thelwall M, eds. *Springer Handbook of Science and Technology Indicators.* Springer, Heidelberg, Germany, 2019: 929–56.
12. Woo H-G, Yeom J, Lee C. Screening early stage ideas in technology development processes: a text mining and k-nearest neighbours approach using patent information. *Technology Analysis & Strategic Management* 2019; **31**: 532–45.
13. Wang HC, Chi YC, Hsin PL. Constructing Patent Maps Using Text Mining to Sustainably Detect Potential Technological Opportunities. *Sustainability* 2018; **10**(10): 3729.
14. Djellabi R, Giannantonio R, Falletta E, Bianchi CL. SWOT analysis of photocatalytic materials towards large scale environmental remediation. *Current Opinion in Chemical Engineering* 2021; **33**: 100696.
15. Chi Y, Wang W, Zhang Q *et al.* Evaluation of practical application potential of a photocatalyst: Ultimate apparent photocatalytic activity. *Chemosphere* 2021; **285**: 131323.

16. Luan C, Song B. Chinese university patenting, patent commercialization and economic growth: A provincial-level analysis. *COLLNET Journal of Scientometrics and Information Management* 2019; **13**: 291–312.
17. Chen Z, Zhang J. Types of patents and driving forces behind the patent growth in China. *Economic Modelling* 2019; **80**: 294–302.
18. She M, Wang Y, Yang X. Antecedents and consequences of strategic patenting for legitimacy: Evidence from China. *Journal of Small Business Management* 2020; **58**: 572–616.
19. Potekhina A, Blind K. What motivates the engineers to patent? A study at the Chinese R&D laboratories of a European MNC. *The Journal of Technology Transfer* 2020; **45**: 461–80.
20. Li Y, Phelps NA, Liu Z, Ma H. The landscape of Chinese invention patents: Quantity, density, and intensity. *Environment and Planning A: Economy and Space* 2019; **51**: 823–6.
21. Yamashita N, Yamauchi I. Innovation responses of Japanese firms to Chinese import competition. *The World Economy* 2020; **43**: 60–80.
22. Lin J, Wu H-M, Wu H. Could government lead the way? Evaluation of China's patent subsidy policy on patent quality. *China Economic Review* 2021; **69**: 101663.
23. Hsu DH, Hsu P-H, Zhao Q. Rich on paper? Chinese firms' academic publications, patents, and market value. *Research Policy* 2021; **50**: 104319.
24. Lee JW, Lee WK, Sohn SY. Patenting trends in biometric technology of the Big Five patent offices. *World Patent Information* 2021; **65**: 102040.
25. Jiang R, Shi H, Jefferson GH. Measuring China's International Technology Catchup. *Journal of Contemporary China* 2020; **29**: 519–34.
26. Wang X, Ji G, Zhang Y, Guo Y, Zhao J. Research on High- and Low-Temperature Characteristics of Bitumen Blended with Waste Eggshell Powder. *Materials* 2021; **14**(8): 2020.
27. Quinn G. PCT Basics: Obtaining Patent Rights Around the World. 2021. https://www.ipwatchdog.com/2021/02/03/pct-basics-obtaining-patent-rights-around-world-2/ (accessed August 4, 2021).

doi: 10.2166/9781789061932_0183

Chapter 7
Analysis of patents in photocatalytic water and wastewater treatment. Part II – solar energy and nanotechnology

Ali Zebardasti[1], Mohamad Hosein Nikfar[2], Danilo H. S. Santos[3,4], Lucas Meili[3], Mohammad G. Dekamin[1] and Alireza Bazargan[5]*

[1]Pharmaceutical and Heterocyclic Compounds Research Laboratory, Department of Chemistry, Iran University of Science and Technology, Tehran 1684613114, Iran
[2]Department of Civil Engineering, K. N. Toosi University of Technology, Tehran, Iran
[3]Laboratory of Processes (LAPRO), Center of Technology, Federal University of Alagoas, 57072-970 Maceio, Alagoas, Brazil
[4]Laboratory of Applied Electrochemistry (LEAp), Institute of Chemistry and Biotechnology, Federal University of Alagoas, 57072-970 Maceio, Alagoas, Brazil
[5]School of Environment, College of Engineering, University of Tehran, Tehran, Iran
*Corresponding author: alireza.bazargan@ut.ac.ir

ABSTRACT

The insight provided in the previous chapter, revealed that solar photocatalytic systems as well as the use of nanotechnology in photocatalysis are two of the topics that have received a lot of attention in the patents for photocatalytic water and wastewater treatment. Herein, the most important International Patent Code (IPC) and Cooperative Patent Classification (CPC) patent codes and key concepts with regard to these subtopics are presented, and data regarding the activity of various countries, research institutes and companies is offered. The top patent registrants in solar photocatalysis are from China, Japan, and the US, in that order, and the top patent destination countries follow the same order. Several Chinese universities are the top patent owners of patents in this field, and Jiangsu University emerged as the top owner in this group. The results of text mining showed that titanium dioxide is the most widely used material. As for nanotechnology patents in the field, Chinese patent registrants are leading the field by far. When analyzing the patents on nanotechnology, in contrast to the other topics/subtopics, it was found that research institutions play a significantly greater role than companies.

© 2022 The Editors. This is an Open Access book chapter distributed under the terms of the Creative Commons Attribution Licence (CC BY-NC-ND 4.0), which permits copying and redistribution for noncommercial purposes with no derivatives, provided the original work is properly cited (https://creativecommons.org/licenses/by-nc-nd/4.0/). This does not affect the rights licensed or assigned from any third party in this book. The chapter is from the book Photocatalytic Water and Wastewater Treatment, Alireza Barzagan (Ed.).

7.1 INTRODUCTION TO SOLAR PHOTOCATALYTIC PATENTS

In theory the energy coming from the sun is sufficient to feed all the energy needs of humanity. In fact, the solar energy striking the earth in one hour covers the annual need of humans [1]. Yet, there are many barriers to rapid growth in the field of solar technologies, including technical barriers such as the low efficiency of such systems, and economic and institutional obstacles such as lack of financing mechanisms, policy and infrastructure limitations, and a shortage of skilled workers [2]. Furthermore, it is misleading just to speak about the amount of solar energy hitting the earth. Studies have shown that the constrained solar potential worldwide (the amount which is deliverable to end-users after subtracting the energy inputs needed for capital infrastructure and operation) is a little over 1000 exajoules per year, of which 98%, 75%, and only 15% can be extracted if the system needs to deliver an energy return on energy invested set at 5, 7.5, and 9, respectively. This is a much lower solar power potential than had been previously estimated. The achievable potential will be greatly constrained by how high the energy return needs to be in comparison with the energy required to maintain a sustainable society. In regions with low solar radiation, the effect is especially pronounced [1].

In recent years a lot of research has been focused on photocatalysis as one way of converting solar energy into chemical energy. In addition to decomposing organic/inorganic pollutants, inactivating bacteria, reducing CO_2, and reducing N_2 by photocatalysis, there are many other applications of solar photocatalysis. Using solar light as the energy source for photocatalysis is one of the holy grails in the field, and many studies have been published that relate to solar light-driven photocatalysis [3]. In order to widen the solar light response and accelerate charge migration, the use of heterostructure-based full-solar-light-driven photocatalysts could be a sensible approach [4].

Figure 7.1 shows the number of patents published in solar photocatalysis for water and wastewater treatment. As evident, the registration of patents in this area began in the late 80s and continued without much change until the mid-90s. From then until 2009 an increase in patent registrations per year was observed. Following 2009 there was a boom until 2012 reaching nearly 120 patents per year. After a significant fall in 2013, the number of patents again grew with a sharp increase until 2019. The data in Figure 7.1 was extracted in mid-2020 and there-fore the number of patents filed in 2020 is not complete. With some exceptions, the overall trend of patent registrations in solar photocatalytic water and wastewater treatment systems has been increasing. From Figure 7.2, it can be seen that the same is true regarding peer-reviewed academic papers on photocatalysis in general, and solar photocatalysis specifically. Evidently, solar photocatalysis is only a subtopic of photocatalysis, which includes many other subjects and subtopics.

Solar photocatalysis is especially interesting because in addition to making photocatalytic water and wastewater treatment more accessible due to lower investment and reduced energy consumption [5] the development of appropriate technologies that integrate solar energy into water treatment processes can contribute to ending the world's serious water shortage. This will be helped by the fact that, coincidentally, many of the arid and semi-arid regions of the world that face water scarcity are blessed with abundant solar radiation [6–8].

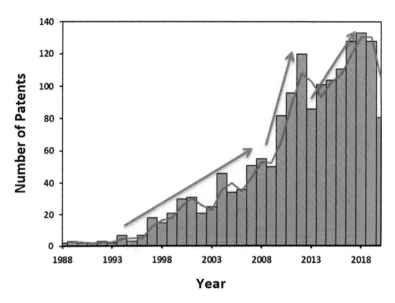

Figure 7.1 The number of filed patents per year in solar photocatalytic water and wastewater treatment.

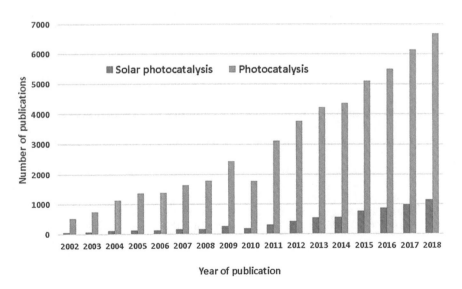

Figure 7.2 The publication of papers on photocatalysis as well as solar photocatalysis per year. Reproduced from [9].

7.2 ANALYSIS OF SOLAR PHOTOCATALYTIC PATENTS

In order to carry out patent analysis it is very important to first identify the relevant standard codes which need to be searched. As discussed in the previous chapter, keyword searches and extraction of patents through software can be useful in this regard.

In terms of International Patent Code (IPC), a total of 19 main codes were found to be relevant, the most important of which are listed in Table 7.1. The percentage shown in Table 7.1 identifies what fraction of the patents were found in each code. Evidently, code C02F1/00 is the most popular main code with roughly 29% of the patents. Table 7.1 also shows the main subgroups in the IPC codes, the most popular of which is C02F1/30 with 22% of the patents. Since C02F1/30 is a subgroup of C02F1/00 this implies that as per the IPC the largest portion of the patents is filed under irradiation for water treatment purposes. From among the IPC main codes, nearly all the top codes are identical to those for photocatalytic materials for water and wastewater treatment (presented in the previous chapter). The only exceptions are the appearance of codes C02F3/00 (Biological treatment of water) and B01J19/00 (Chemical) further up on the list. Other codes have only slightly shifted in their rank. The IPC subgroups are even more similar than the main groups, with all top subgroups being the same and only some slight rank changes further down the table.

On the other hand, Table 7.2 shows the top standard codes for relevant patents as per the Cooperative Patent Classification (CPC) system which has been developed by the European Patent Office with the cooperation of the US Patent and Trademark Office. Here the largest proportion of relevant patents, about 35%, are found under the main group code Y02W10/00 (Technologies for wastewater treatment). From the subgroups of the CPC, the most prolific code was found to be Y02W10/37 (using solar energy) which is a logical outcome of the search. As for the similarity of the main CPC codes with those found for photocatalytic materials in the previous chapter, the majority of the top main CPC codes are the same with some new entrants such as Y02E10/00 (Energy generation through renewable energy sources).

Although solar photocatalytic water treatment is a global concern, the intensity of research in this area may vary in different countries. Based on the analysis of the number of patents per country of origin presented in Figure 7.3, the Chinese are the most active in the filing of patents on solar photocatalytic treatment. China (CN) researchers contributed 806 patents. This result is understandable given the significant efforts of the Chinese government dedicated to resolving water supply and demand conflicts in the country [10]. Patent filing by the Japanese (JP), Americans (US) and South Koreans (KR) come next with 212, 97 and 46 patents, respectively. Although in smaller numbers, Europeans from France (FR) and Italy (IT) are also among the top 15 in this area.

Figure 7.4 shows the 11 main destination countries for patent registration. The largest number of patents are registered in China, with a significant 1103 (67.3%) portion of the total, followed by Japan (JP), the United States (US), WIPO (WO, World Intellectual Property Organization), South Korea (KR) and EPO (EP, European Patent Office) with 230 (14%), 165 (10%), 52 (3.17%), 28 (1.7%) and 14 (0.85%) patents, respectively.

Photocatalytic patents with solar and nanotechnology 187

Table 7.1 The most frequent IPC patent codes used for patents relevant to solar photocatalytic water and wastewater treatment.

	Code Used	% of Patents	Description
Main groups	C02F1/00	29	Treatment of water
	B01J35/00	11	Catalysts
	B01J23/00	8	Catalysts comprising metals or metal oxides or hydroxides
	C02F101/00	7	Nature of the contaminant
	B01J21/00	7	Catalysts comprising the elements
	B01J27/00	7	Catalysts comprising the elements or compounds of halogens
	C02F9/00	5	Multistep treatment of water
	B01J37/00	4	Processes
	B01D53/00	3	Separation of gases or vapors
	C01B3/00	3	Hydrogen
	B01J31/00	2	Catalysts comprising hydrides
	C02F3/00	2	Biological treatment of water
	C02F103/00	2	Nature of water
	C01G23/00	2	Compounds of titanium
	B01J19/00	2	Chemical
Main subgroups	C02F1/30	22	By irradiation
	C02F1/32	15	With ultra-violet light
	C02F1/72	14	By oxidation
	B01J35/02	9	Solids
	B01J21/06	7	Silicon
	C02F101/30	4	Organic compounds
	C01B3/04	3	By decomposition of inorganic compounds
	C02F101/38	3	Containing nitrogen
	B01D53/86	3	Catalytic processes
	C02F1/28	3	By sorption
	B01J27/24	2	Nitrogen compounds
	C02F9/08	2	At least one step being a physical treatment
	C02F101/34	2	Containing oxygen
	B01J35/10	2	Characterized by their surface properties or porosity
	C02F9/14	2	At least one step being a biological treatment

As explained earlier, the European Patent Office receives more applications than national offices of member states because applicants may apply via the EPO to seek protection within all EPO member states. An application through the Patent Cooperation Treaty designated with a WO prefix standing in fourth place in Figure 7.4 is also of importance. This means that other than patents

Table 7.2 The most frequent CPC patent codes used for patents relevant to solar photocatalytic water and wastewater treatment.

	Code Used	% of Patents	Description
Main groups	Y02W10/00	35	Technologies for wastewater treatment
	C02F1/00	10	Treatment of water
	C02F2305/00	8	Use of specific compounds during water treatment
	B01J35/00	6	Catalysts
	Y02E60/00	5	Enabling technologies or technologies with potential or indirect contribution to GHG emissions mitigation
	C02F2101/00	4	Nature of the contaminant
	B01J37/00	4	Processes
	B01J21/00	3	Catalysts comprising the elements
	Y02E10/00	3	Energy generation through renewable energy sources
	C02F2201/00	3	Apparatus for treatment of water
	B01J23/00	3	Catalysts comprising metals or metal oxides or hydroxides
	Y02A20/00	2	Water conservation; efficient water supply; efficient water use
	Y02P20/00	2	Technologies relating to chemical industry
Main subgroups	Y02W10/37	38	Using solar energy
	C02F2305/10	8	Photocatalysts
	C02F1/725	8	By catalytic oxidation
	B01J35/004	7	Photocatalysts
	C02F1/32	6	With ultra-violet light
	Y02E60/36	5	Hydrogen production from non-carbon containing sources
	C02F1/30	3	By irradiation
	C02F1/325	3	Irradiation devices or lamp constructions
	B01J21/063	3	Titanium; oxides or hydroxides thereof
	Y02A20/212	2	Solar powered wastewater sewage treatment
	Y02P20/133	2	Renewable energy sources
	B01J35/002	2	Catalysts characterized by their physical properties
	B82Y30/00	2	Nanotechnology for materials or surface science
	C02F2303/04	2	Disinfection

Figure 7.3 Total number of filed patents in the field of photocatalytic water and wastewater treatment with solar energy by owners from each country as of mid-2020. CN: China; JP: Japan; US: United States; KR: South Korea; DE: Germany; TW: Chinese Taipei (Taiwan); FR: France; SA: Saudi Arabia; CA: Canada; IT: Italy; SG: Singapore; IN: India; AU: Australia; GB: Great Britain; ES: Spain.

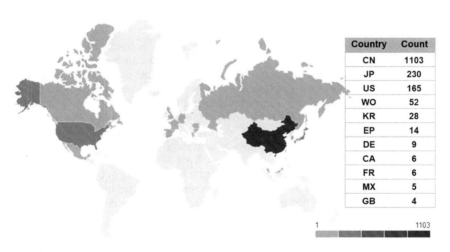

Figure 7.4 Total number of patents filed in the field of solar photocatalytic water and wastewater treatment in each country or territory as of mid-2020. CN: China; JP: Japan; US: United States; WO: Patent Cooperation Treaty (PCT) also known as International Application; KR: South Korea; EP: European Patent Office; DE: Germany; CA: Canada; FR: France; MX: Mexico; GB: Great Britain.

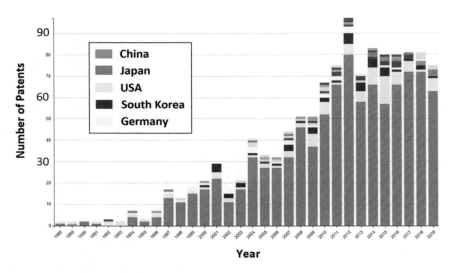

Figure 7.5 Patents filed in solar photocatalytic water and wastewater treatment by nationality of patent holder. China: green; Japan: red; USA: light blue; Korea: dark blue; Germany: yellow.

filed in China, Japan, and the USA, the preference of those who file patents is to go through the PCT process rather than directly going to the patent office of the registering country.

The consolidation of the Chinese as the largest holders of patents in this topic area has occurred in the past 10 years or so, as can be seen in Figure 7.5. Japanese researchers were pioneers in the development of this technology from 1994–2000, however perhaps due to the worsening of water scarcity in China caused by serious problems of contamination [11], as well as rapid economic growth and the large number of industrial production activities in China, its investments in research dealing with the use of advanced technology for water treatment (including solar photocatalysis) have increased making China the top player. The active policy inside China for the development of water treatment technologies is even more evident when looking at the trend of patent registration by destination (Figure 7.6). In summary, the most patents in the world on the topic are not only filed by the Chinese, but are also filed inside their own country.

The analysis of patent holders showed almost an even split between universities (research institutions) and private companies, although the universities edged ahead of the companies 51.7% to 47.8%. The remaining 0.5% of the patents are attributed to individuals and/or have unidentified origins.

An analysis of the profile of the 15 universities with the highest number of filed patents in solar photocatalytic water and wastewater treatment is presented in Table 7.3. The Chinese Academy of Nanjing stands on top with the greatest number of patents in the field at 24 documents. Jiangsu University and Shanghai Jiaotong University follow, with 23 and 18 patents, respectively.

Photocatalytic patents with solar and nanotechnology

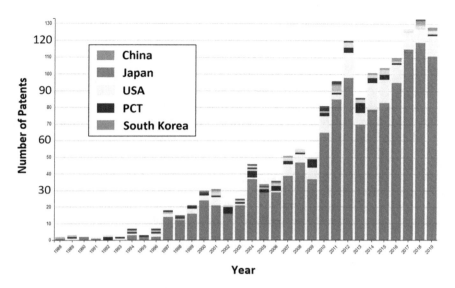

Figure 7.6 Patents filed in solar photocatalytic water and wastewater treatment by destination country. China: green; Japan: red; USA: yellow; World Intellectual Property Organization (PCT): dark blue; South Korea: cyan.

Table 7.3 List of top research institutes and universities filing patents in the field of photocatalytic water treatment using solar energy.

Research Institute or University	Number of Filed Patents
University of Nanjing	24
University of Jiangsu	23
University of Shanghai Jiaotong	18
University of Shandong	14
University of Shanghai	14
University of Henan Normal	14
University of Tongji	12
University of Fuzhou	12
University of Hohai	12
University of South China Tech	11
University of Nanchang Hangkong	11
University of Beijing Normal	10
University of Zhejiang	10
University of Nankai	9
University of Taiyuan Technology	9

There is a notable domination of Chinese universities in the list of research institutions, with no other countries present in the top 15 positions. A closer examination of the temporal activity of universities in this regard shows that Jiangsu University has had considerable activity in recent years, with a couple of new patents being filed at the university in solar photocatalysis every year since 2013.

An analysis of the profile of the companies that have filed the most patents is presented in Figure 7.7. The Japanese Ebara Corporation, a manufacturer of industrial machines, is the forerunner in the registration of patents in solar photocatalysis for water and wastewater treatment, having filed 13 patents. They are followed by Kanagawa Kagaku Gijutsu Akad, Babcock Hitachi KK, and Sharp KK companies, also of Japanese origin, with eight patents each. The Chinese company Chengdu New Keli Chem Sci co, is also a prominent patent holder with eight patents, but further scrutiny of the data has shown that this company does not have a long history in patenting photocatalytic technologies,

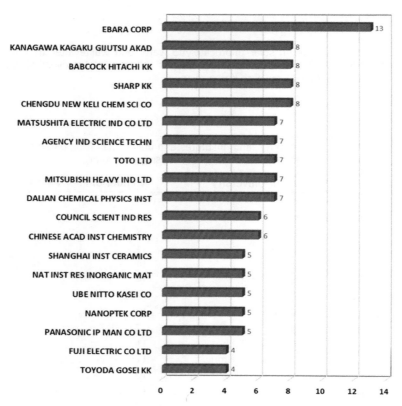

Figure 7.7 List of companies with the largest number of filed patents in photocatalytic water and wastewater purification using solar energy.

Photocatalytic patents with solar and nanotechnology

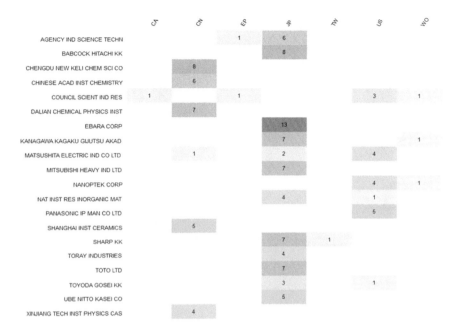

Figure 7.8 Main destination for patent registration by the top players in the field.

with all of its eight patents being filed in the same year, 2018. Unlike what is observed in the list of the top universities in registering patents, where Chinese hegemony is unrivaled, the domain of patents belonging to companies is mostly occupied by the Japanese.

Figure 7.8 shows the destination countries for patents registration for the top players in the field. As indicated, the main destinations for patent registration by active companies are Japan, China and the United States, in that order.

Patent registration cooperation between companies is shown in Figure 7.9. It is possible for companies to cooperate in order to achieve patentable knowledge together. This will require collaborative research and development and synchronized management. However, companies that seek monopolization of their technologies, limit their collaborations in general because they seek to maximize the exclusivity of their technology. Keeping trade secrets is another way to retain competitive advantage and prevent competitors from finding out what is known. According to the figure, most companies have kept their list of partners limited, but Kanagawa Kagaku Gijutsu Akad has shown the highest number of collaborations, registering joint patents with five other companies.

One of the most important results of patent analysis is extracting key concepts from the patents. Based on the extracted information, items such as technology subdisciplines, materials used, the application of the technology and many other issues can be discerned. Based on the analysis performed on solar photocatalytic water and wastewater treatment, the main concepts (words) were extracted and are presented in Table 7.4.

194 Photocatalytic Water and Wastewater Treatment

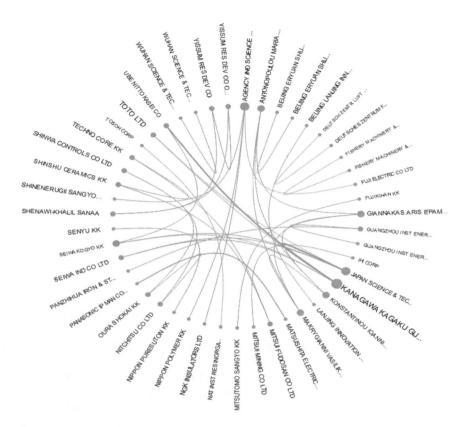

Figure 7.9 Cooperation between companies for patent registration.

Table 7.4 shows the top 50 most frequently repeated concepts/words in this field. The list of words mainly points to the materials used in the fabrication of photocatalysts, and more specifically, the list is dominated by titanium dioxide and its various combinations with other words. In fact, titanium appears in 15 of the rows in Table 7.4. According to the results, the four most cited terms were: titanium dioxide, energy consumption, carbon dioxide, and valence band. Titanium dioxide (TiO$_2$) is known for its efficiency in photocatalytic processes when used alone or with other compounds such as graphene to minimize the charge recombination rate of the photogenerated electron and increase the visible light absorption capacity of the TiO$_2$, and consequently the efficiency of the process [12]. Titanium dioxide can undergo various modifications to increase its performance as a photocatalyst, including doping, thin film immobilization, and composite fabrication. In the academic literature, the use of TiO$_2$ for degrading special classes of contaminants, such as those now referred to as 'emerging pollutants' [13] or 'humic acid' which is the main component of landfill leachate, has received ample attention [14]. TiO$_2$ reduced

Photocatalytic patents with solar and nanotechnology 195

Table 7.4 The top 50 most frequent words seen in patents on photocatalytic water treatment using solar energy.

Rank	Word(s)	Number of Patents	Rank	Word(s)	Number of Patents
1	Titanium dioxide	351	26	Titanium tetrachloride	39
2	Energy consumption	131	27	Quartz glass	39
3	Carbon dioxide	113	28	Cadmium sulfide	39
4	Valence band	110	29	Carbon nanotubes	38
5	Hydrogen peroxide	107	30	Xenon lamp	36
6	Hydroxyl radicals	90	31	Titanium dioxide nano	35
7	Mercury lamp	85	32	TiO_2 photocatalysis	34
8	Ultraviolet lamp	72	33	Nanometer titanium dioxide	32
9	Ethyl alcohol	71	34	Chemical energy	31
10	Reaction kettle	67	35	Silicon dioxide	31
11	Sodium hydroxide	66	36	TiO_2 nanoparticles	30
12	Methylene blue	64	37	Redox reactions	30
13	Ultra violet rays	62	38	Reaction tank	30
14	TiO_2 photocatalyst	62	39	Titanium dioxide powder	29
15	Methyl orange	58	40	Titanium dioxide particles	29
16	Photocatalyst particles	50	41	Nano TiO_2	28
17	Chemical reaction	50	42	Adsorption capacity	28
18	Tetrabutyl titanate	48	43	Catalyst particles	27
19	Absorption spectrum	47	44	Polyethylene glycol	27
20	Ethylene glycol	45	45	TiO_2 powder	26
21	TiO_2 particles	42	46	Chemical substance	26
22	Sol Gel	40	47	Polyvinyl alcohol	26
23	Titanium dioxide photocatalyst	40	48	Photocatalysis sewage	26
24	Reaction vessel	39	49	Oxygen species	26
25	Energy conversion	39	50	Titanium dioxide nanotube	25

by removing oxygen atoms or by adding hydrogen has been proven to be useful for a variety of photocatalytic tasks such as photodegradation of organic compounds, hydrogen generation from water splitting, and CO_2 reduction for CH_4 evolution amongst others. Further enhancements of TiO_2-x are possible by self-doping with Ti^{3+} and/or with other traditional modifications [15]. Titanium dioxide may be doped with metals or non-metals [16]. According to one review, annealing and hydrothermal methods were more efficient than

196 Photocatalytic Water and Wastewater Treatment

others for preparing non-metal doped photocatalysts. Both methods can be used successfully, but hydrothermal methods can use a variety of precursors while annealing methods can only be used on solid precursors. Thus, it was concluded that the hydrothermal method is more favorable [17]. The frequent mention of nanotechnology alongside titanium dioxide warrants the separate scrutiny of nanotech in photocatalytic water and wastewater treatment, which will be discussed in the second half of this chapter which follows.

7.3 NANOTECHNOLOGY IN PHOTOCATALYTIC TREATMENT OF WATER AND WASTEWATER

In this section, an analysis of patents regarding nanotechnology in photocatalytic water and wastewater treatment is provided. After searching the databases and extracting the related patents, a total of 3610 related patents were identified. This is the number of filed patents that have been analyzed, and includes both patents that have been granted and those that have been rejected.

Nanomaterials have unique size-dependent properties that create a large potential for the removal of contaminants [18]. A relevant aspect of nanotechnology in photocatalysis is the fabrication of nanophotocatalysts or the nano-sized modification of larger photocatalytic particles which can dramatically improve their photocatalytic activity [19]. The nano-sized metal oxides have a significant 'size quantification' effect as a result of their high surface to mass ratio, resulting from their small size. The increase in surface area can, however, lead to poor stability as the size shrinks to nanometer levels. Agglomeration is therefore a limitation of nanomaterials due to Van der Waals forces and other interactions. Often, these features are determined by morphology, which is controlled by synthesis conditions. Aqueous contaminants can be degraded by photocatalysis in two stages, adsorption and degradation. Adsorption can occur without light, but degradation only occurs when the photocatalyst is exposed to light [20].

Two-dimensional (2D) nanosheet photocatalysts have many advantages over traditional three-dimensional nanopowder photocatalysts. Among these benefits are improved light adsorption characteristics, and shorter electron and hole migration paths to the photocatalyst's surface, which minimize undesirable electron-hole pair recombination. It is possible to modulate the band gap of 2D materials and transfer charges from the semiconductor to the adsorbates through surface defects. If they are exploited together and optimized, these factors can confer remarkable activities on 2D photocatalysts relative to their 3D counterparts. As a result, a wide range of experimental approaches are being explored for the synthesis of 2D photocatalysts, while computational approaches are increasingly being used to identify promising new 2D photocatalysts [21]. In addition to applications such as photocatalytic water disinfection applications, the nanostructures can be used as building blocks for other purposes. There are numerous types of 2D photocatalytic materials, some of which are graphene, graphitic carbon nitride, 2D metal oxides and metallates, metal oxyhalides and transition metal dichalcogenides [22–24]. Smaller still, one-dimensional (1D) and zero-dimensional (0D) nanomaterials have attracted great attention in the

Photocatalytic patents with solar and nanotechnology

field of photocatalysis. Unfortunately, they still face many obstacles, such as electron-hole pair recombination and low apparent quantum efficiency, which result in poor photocatalytic performance in practical applications. In order to address these challenges, photocatalytic heterojunctions have been constructed using 1D nanomaterials as building blocks, and their characteristics have been demonstrated to be unique. Types of 1D nanomaterial-based heterojunctions include the type II heterojunction, p–n type heterojunction, Schottky junction, Z-type heterojunction, and S-scheme heterojunction [25].

To perform a patent analysis, the relevant standard codes must be identified first so that they can be searched. In the previous chapter it was discussed that keyword searches and using software to extract information from patents could prove useful in this regard. The most relevant IPC and CPC codes are displayed in Tables 7.5 and 7.6. The most notable IPC main groups which had not appeared earlier in this chapter or in the previous chapter are the 'Manufacture or treatment of nanostructures' (B82Y40/00) and 'Nanotechnology for materials or surface science' (B82Y30/00) codes.

Figure 7.10 shows the temporal trend of patent registration for photocatalytic water and wastewater treatment with nanotechnology. As can be seen, the registration of patents started in 1989, but it took more than a decade for pace to pick up. According to the data, the registration process can be divided into three general periods. In the first period comprising the 90s there was little to no activity. In the second period, from 1999 to the 2010s, there was a considerable increase in the number of patent registrations. However, in the third period from 2015 onward, the number of patent registrations considerably increased, surpassing 400 patents a year in 2018. The data for the year 2020 is not indicative of a decrease because the current analysis was carried out in mid-2020.

Based on the analysis of the nationality of patent registrants shown in Figure 7.11, it is evident that the Chinese (CN) are the most active by far, with 1998 patents. The Japanese (JP) rank at a very distant second with 151 patents closely followed by the United States (US) with 150 patents. Although on a much smaller scale, registrants from European countries such as France (FR) and Germany (DE), also appear on the list. The high number of patent filings by registrants from Asian countries comprising 4 of the top 5, may be the result of the sizable investments made by these countries in research and development in recent years. It should also be noted that country-level data does not necessarily provide an accurate picture [26]. For example, a study of Chinese science and technology investment has shown that regions within the country have large differences in investment, with Eastern China being the barycenter of the Chinese economic map and one of the fastest economically developing areas in the world [27]. In addition, obviously, not all registered patents are of equal worth. Recent years have seen an explosion of scientific output and unfortunately scientific misconduct as well; therefore the Chinese government has been trying to curb unprofessional academic and scientific behavior to rectify transgressions where they may exist [28]. Figure 7.12 shows the destination countries for patent registration. As with all other cases investigated in the book, the highest number of patents by far are registered in

198 Photocatalytic Water and Wastewater Treatment

Table 7.5 The most frequent IPC patent codes for photocatalytic water and wastewater treatment using nanotechnology.

	Code Used	% of Patents	Description
Main groups	C02F1/00	20	Treatment of water
	B01J35/00	9	Catalysts
	B01J23/00	9	Catalysts comprising metals or metal oxides or hydroxides
	C02F101/00	9	Nature of the contaminant
	B01J21/00	8	Catalysts comprising the elements
	B01J27/00	8	Catalysts comprising the elements or compounds of halogens
	B01J37/00	6	Processes
	C02F9/00	4	Multistep treatment of water
	B01J31/00	3	Catalysts comprising hydrides
	B01D53/00	3	Separation of gases or vapors
	B82Y40/00	3	Manufacture or treatment of nanostructures
	C01B3/00	3	Hydrogen
	C01G23/00	3	Compounds of titanium
	B82Y30/00	3	Nanotechnology for materials or surface science
Main subgroups	C02F1/30	18	By irradiation
	C02F1/32	10	With ultra-violet light
	B01J21/06	9	Silicon
	C02F1/72	7	By oxidation
	C02F101/30	6	Organic compounds
	B01J35/02	6	Solids
	C02F101/38	4	Containing nitrogen
	B01J27/24	4	Nitrogen compounds
	B01D53/86	4	Catalytic processes
	C01B3/04	3	By decomposition of inorganic compounds
	B01J35/10	3	Characterized by their surface properties or porosity
	C02F101/34	3	Containing oxygen

China, with 2927 patents. The United States is the next target with 317 filed patents. The low number of patents filed at the European Patent Office in this subtopic of photocatalysis is noteworthy.

Overall, the patent analysis showed a complete dominance of Chinese registrants and China as the destination market of the patents in terms of numbers. However, one must approach these findings with caution, because the sheer number of patents is not an indication of a technology's applicability or usefulness. Studies have shown that the number of patent filings or citations

Photocatalytic patents with solar and nanotechnology 199

Table 7.6 The most frequent CPC patent codes for photocatalytic water and wastewater treatment using nanotechnology.

	Code Used	% of Patents	Description
Main groups	Y02W10/00	12	Technologies for wastewater treatment
	B01J35/00	11	Catalysts
	C02F1/00	10	Treatment of water
	C02F2305/00	8	Use of specific compounds during water treatment
	B01J37/00	7	Processes
	B01J21/00	7	Catalysts comprising the elements
	B01J23/00	5	Catalysts comprising metals or metal oxides or hydroxides
	Y02E60/00	4	Enabling technologies or technologies with potential or indirect contribution to GHG emissions mitigation
	B82Y30/00	4	Nanotechnology for materials or surface science
	C02F2101/00	4	Nature of the contaminant
	C01P2004/00	4	Particle morphology
	C01G23/00	3	Compounds of titanium
	B01D53/00	3	Separation of gases or vapors; Recovering vapors or volatile solvents from gases; Chemical or biological purification of waste gases
Main subgroups	B01J35/004	13	Photocatalysts
	Y02W10/37	13	Using solar energy
	C02F2305/10	10	Photocatalysts
	C02F1/725	8	By catalytic oxidation
	B01J21/063	7	Titanium; oxides or hydroxides thereof
	C02F1/32	7	With ultra-violet light
	Y02E60/36	4	Hydrogen production from non-carbon containing sources
	B01J35/0013	4	Colloids
	C02F1/30	4	By irradiation
	C02F1/325	3	Irradiation devices or lamp constructions
	B01J35/002	3	Catalysts characterized by their physical properties
	C01P2004/64	3	Nanometer sized
	B01D2255/802	3	Photocatalytic

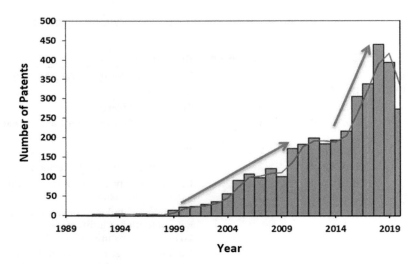

Figure 7.10 Patent registrations per year in photocatalytic water and wastewater treatment with nanotechnology.

filed by a company does not represent the success or failure of its R&D [29]. However, patent counts can be a reflection of a firm's product and process innovation mix. Intuitively, it is difficult to detect competitor infringements of innovative processes, which suggests these innovations are better protected by trade secrets than by patents [29]. In fact, as numerous studies have shown,

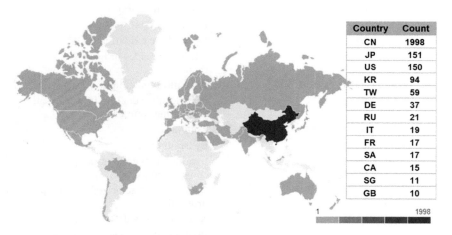

Figure 7.11 The country of origin for registrants of patents in the area of nanotechnology for photocatalytic water and wastewater treatment as of mid-2020. CN: China; JP: Japan; US: United States; KR: South Korea; TW: Chinese Taipei (Taiwan); DE: Germany; RU: Russia; IT: Italy; FR: France; SA: Saudi Arabia; CA: Canada; SG: Singapore; GB: Great Britain.

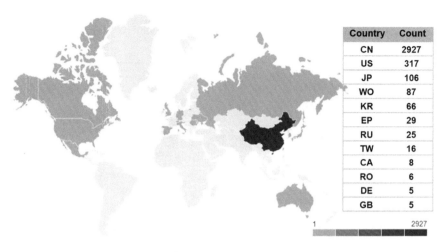

Figure 7.12 Destination countries for patents on nanotechnology in photocatalytic water and wastewater treatment as of mid-2020. CN: China; US: United States; JP: Japan; WO: Patent Cooperation Treaty (PCT) also known as International Application; KR: South Korea; EP: European Patent Office; RU: Russia; TW: Chinese Taipei (Taiwan); CA: Canada; RO: Romania; DE: Germany; GB: Great Britain.

many advancements across various sectors are kept as trade secrets and not published, neither as academic papers nor patents [30–33]. Trade secrets have often been cited as the most important means of appropriating knowledge and the basis for competitive advantage by managers, even more so than formal intellectual property protection, despite the fact that it is very difficult to quantify and study such secrets [34]. In terms of commercial strategies, patenting offers the greatest variety, not only of in-house production and marketing, but also of patent sales and different types of licensing arrangements. In addition, there are hybrid options, such as combining patenting with free licensing, which has characteristics similar to academic journal publishing. However, this will entail additional costs – both direct and indirect – of patenting. Even though patenting, publishing, and secrecy are commonly used as substitutes for one invention, it is essential to consider the combinatorial opportunities across time and across multiple inventions and technologies. An invention that becomes a patent must always go through some kind of secrecy to ensure that it is novel and therefore patentable. Different strategies can also be combined for related, but distinct inventions. For example, whenever there are multiple modules in a product system, it is possible to protect one module with trade secrets, while another module might be protected by a patent, with information regarding the third module published to create low-priced substitutes or complements for the entire system [35].

The results of the patent analysis show that 1907 patents (52.8% of the total) were registered by academic research institutions, while 1449 patents (40.1% of the total), belong to companies, and the remaining 7.1% are attributed

Table 7.7 List of the top 16 universities that have filed patents in photocatalytic water and wastewater treatment with nanotechnology.

Academic	Count
University of Jiangsu	58
University of Zhejiang	40
University of Shanghai	37
University of Changzhou	36
University of Shanghai Jiaotong	31
University of Tianjin	30
University of Nanjing	28
University of Fuzhou	25
University of Tongji	23
University of Hohai	23
University of Hunan	22
University of Shaanxi Science & Tech	22
University of Wuhan Tech	21
University of Shandong	21
University of Jinan	21
University of Ningbo	21

to individuals and/or have unidentified origin. Thus, again, it seems that universities have a slight edge over other registrants. A list of the top 16 most prolific universities is shown in Table 7.7. The universities of Jiangsu and Zhejiang, China, emerged as the top active universities in the group, with 58 and 40 registrations, respectively. There is a remarkable dominance of the list by Chinese universities. Furthermore, Figure 7.13 shows the number of patents filed by each university since the turn of the millennium. This figure is interesting because it shows that although Jiangsu University is at the top of

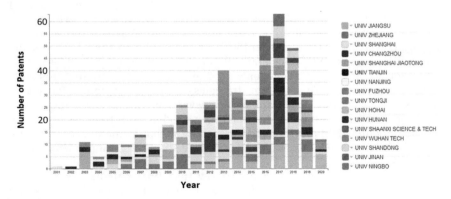

Figure 7.13 Trend of patent registration by university per year.

Table 7.7, it did not have any patent filings on the topic prior to 2010; and since then it has shown significant activity. Meanwhile, although the University of Zhejiang and University of Shanghai rank lower in Table 7.7, they have had more consistent activity for a longer duration of time. Conversely, some institutions have only had a short-lived burst of registrations, the most apparent being the university of Ningbo with many filings in 2013 and very little activity before or after that. As evident, the most fruitful year overall for universities was 2017 with a combined filing of more than 60 patents.

An analysis of companies filing the most patents on the subject is presented in Table 7.8. The Japanese company TOTO LTD is a forerunner in the registration of patents in nanotechnology in photocatalytic water and wastewater treatment, having 16 patents. Chinese company Chengdu New Keli Chem Sci co is the next most prolific company in this field, with 13 filings. After that a group of companies, only one of which is not Asian (Procter & Gamble) have filed 10 patents each. As evident from Figure 7.14, although most other registrants in the list have focused on the Chinese market and have seldom shown activity elsewhere, TOTO LTD and Procter & Gamble have filed seven and six patents in the US, respectively.

Table 7.9 indicates the most common concepts/words of significance in the filed patents. According to these results, the term titanium dioxide emerged as the most used, which is similar to what was seen before. Nano derivatives (nano, nanometer, and nanoparticle) were very frequently used words.

Table 7.8 List of companies with the most filed patents relevant to nanotechnology in photocatalytic water and wastewater treatment.

Company	Count
Toto Ltd	16
Chengdu New Keli Chem Sci Co	13
Chinese Acad Tech Inst Physics	10
Procter & Gamble	10
Hefei Linuo New Mat Trade Co Ltd	10
Anhui Huiming Construction Group Co Ltd	10
Chinese Acad Inst Chemistry	9
Hefei Inst Physical Sci Cas	9
Univ Shanghai Electric Power	9
Shanghai Inst Ceramics	8
Ube Industries	8
Sh Nat Eng Res Ct Nanotech Co	7
Chinese Res Acad Env Sciences	7
Ind Tech Res Inst	6
Changchun Gold Res Inst	6
Anhui Shifang New Glass Technology Co Ltd	6
Henkel Kgaa	5
Gm Global Tech Operations Inc	5

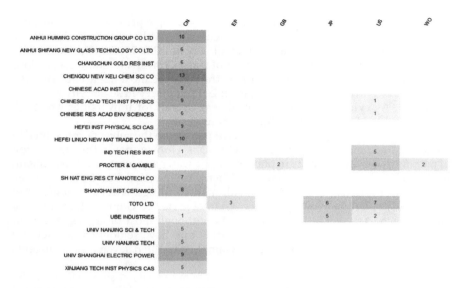

Figure 7.14 Patent filing activity of companies based on target destination.

7.4 CONCLUDING REMARKS

In this chapter patent analysis was carried out for photocatalytic water and wastewater treatment using solar energy, followed by a section devoted to the use of nanotechnology in patents. Overall, by searching databases, a total of 1637 patents were selected for solar energy, and 3610 patents for nanotechnology. This is the number of filed patents, and includes both granted and rejected applications.

For solar energy, the top CPC codes are Y02W10/00, C02F1/00, and B01J35/00. The results of intelligent text mining in this field also showed that titanium dioxide photocatalyst appears as the most widely used material. Moreover, the keywords that were repeatedly extracted from patents included concentration, efficiency, performance, activity, surface area, contaminants, impurities, formaldehyde, methylene, unpleasant odor, and organic components.

For nanotechnology, the top CPC codes are Y02W10/00, B01J35/00, and C02F1/00, with Y02W10/00 being the top code for both subtopics investigated. The results of intelligent text mining in this field also showed that titanium dioxide is the most widely used word. Polymers such as polyethylene glycol and polyvinyl alcohol are also prevalent.

Considering the numbers, the patent analysis showed a complete dominance of Chinese registrants and China as the destination market for patents. Nevertheless, one should approach these findings with caution, since the sheer number of patents does not necessarily indicate value or applicability.

Table 7.9 Most common concepts/words of significance appearing in the filed patents.

Rank	Word(s)	Number of Patents	Rank	Word(s)	Number of Patents
1	Titanium dioxide	879	24	Polyethylene glycol	114
2	Ethyl alcohol	268	25	Ultraviolet rays	113
3	Carbon dioxide	255	26	TiO_2 particles	112
4	Nanometer titanium dioxide	227	27	Sol gel	112
5	Hydrogen peroxide	209	28	Nanometer photocatalyst	111
6	Valence band	207	29	Polyvinyl alcohol	105
7	Energy consumption	185	30	Chemical reaction	102
8	Reaction kettle	180	31	Titanium dioxide powder	101
9	Sodium hydroxide	170	32	Silicon dioxide	99
10	Hydroxyl radicals	169	33	Carbon nanotubes	96
11	Titanium dioxide photocatalyst	164	34	Nano particles	88
12	Nano TiO_2	156	35	Butyl titanate	85
13	Ultraviolet lamp	153	36	Silver nitrate	83
14	Tetrabutyl titanate	149	37	TiO_2 powder	82
15	TiO_2 photocatalyst	136	38	Titanium dioxide nano	79
16	Titanium tetrachloride	132	39	Adsorption capacity	78
17	Nano titanium dioxide	131	40	Agent	78
18	Nanometer TiO_2	130	41	TiO_2 nanoparticles	76
19	Methylene blue	127	42	Absorption spectrum	75
20	Titanium dioxide particles	123	43	Photocatalyst particles	74
21	Mercury lamp	122	44	Reaction vessel	73
22	Methyl orange	121	45	Titanium alkoxide	73
23	Ethylene glycol	120	46	Nanometer photocatalysis	72

REFERENCES

1. Dupont E, Koppelaar R, Jeanmart H. Global available solar energy under physical and energy return on investment constraints. *Applied Energy* 2020; **257**: 113968.
2. Kabir E, Kumar P, Kumar S, Adelodun AA, Kim K-H. Solar energy: Potential and future prospects. *Renewable and Sustainable Energy Reviews* 2018; **82**: 894–900.
3. Meng X, Eluagwule B, Wang M, Wang L, Zhang J. 9 - Solar photocatalysis for environmental remediation. In: Hussain CM, Mishra AK, eds. *Handbook of Smart Photocatalytic Materials: Environment, Energy, Emerging Applications and Sustainability*. Elsevier: Amsterdam. 2020: 183–95.

4. Liu J, Ma N, Wu W, He Q. Recent progress on photocatalytic heterostructures with full solar spectral responses. *Chemical Engineering Journal* 2020; **393**: 124719.
5. Durán A, Monteagudo JM, San Martín I. Operation costs of the solar photocatalytic degradation of pharmaceuticals in water: A mini-review. *Chemosphere* 2018; **211**: 482–8.
6. Zhang Y, Sivakumar M, Yang S, Enever K, Ramezanianpour M. Application of solar energy in water treatment processes: A review. *Desalination* 2018; **428**: 116–45.
7. Mohammed N, Palaniandy P, Shaik F. Pollutants removal from saline water by solar photocatalysis: a review of experimental and theoretical approaches. *International Journal of Environmental Analytical Chemistry* 2021: 1–21.
8. Tony MA. Solar concentration for green environmental remediation opportunity–international review: advances, constraints and their practice in wastewater treatment. *International Journal of Environmental Analytical Chemistry* 2021: 1–33.
9. Fendrich MA, Quaranta A, Orlandi M, Bettonte M, Miotello A. Solar Concentration for Wastewaters Remediation: A Review of Materials and Technologies. *Applied Sciences*, 2019; **9**(1): 118.
10. Zhu Y, Jiang S, Han X *et al*. A Bibliometrics Review of Water Footprint Research in China: 2003–2018. *Sustainability*, 2019; **11**(18): 5082.
11. Wu J. Challenges for Safe and Healthy Drinking Water in China. *Current Environmental Health Reports* 2020; **7**: 292–302.
12. Chen D, Cheng Y, Zhou N *et al*. Photocatalytic degradation of organic pollutants using TiO2-based photocatalysts: A review. *Journal of Cleaner Production* 2020; **268**: 121725.
13. Gopinath KP, Madhav NV, Krishnan A, Malolan R, Rangarajan G. Present applications of titanium dioxide for the photocatalytic removal of pollutants from water: A review. *Journal of Environmental Management* 2020; **270**: 110906.
14. Tung TX, Xu D, Zhang Y, Zhou Q, Wu Z. Removing Humic Acid from Aqueous Solution Using Titanium Dioxide: A Review. *Polish Journal of Environmental Studies* 2019; **28**: 529–42.
15. Fang W, Xing M, Zhang J. Modifications on reduced titanium dioxide photocatalysts: A review. *Journal of Photochemistry and Photobiology C: Photochemistry Reviews* 2017; **32**: 21–39.
16. Li R, Li T, Zhou Q. Impact of Titanium Dioxide (TiO2) Modification on Its Application to Pollution Treatment—A Review. *Catalysts*, 2020; **10**(7): 804.
17. Nasirian M, Lin YP, Bustillo-Lecompte CF, Mehrvar M. Enhancement of photocatalytic activity of titanium dioxide using non-metal doping methods under visible light: a review. *International Journal of Environmental Science and Technology* 2018; **15**: 2009–32.
18. Zhao L, Deng J, Sun P *et al*. Nanomaterials for treating emerging contaminants in water by adsorption and photocatalysis: Systematic review and bibliometric analysis. *Science of The Total Environment* 2018; **627**: 1253–63.
19. Yaqoob AA, Parveen T, Umar K, Mohamad Ibrahim MN. Role of Nanomaterials in the Treatment of Wastewater: A Review. *Water*, 2020; **12**(2): 495.
20. Bishoge OK, Zhang L, Suntu SL, Jin H, Zewde AA, Qi Z. Remediation of water and wastewater by using engineered nanomaterials: A review. *Journal of Environmental Science and Health, Part A* 2018; **53**: 537–54.
21. Zhao Y, Zhang S, Shi R, Waterhouse GIN, Tang J, Zhang T. Two-dimensional photocatalyst design: A critical review of recent experimental and computational advances. *Materials Today* 2020; **34**: 78–91.
22. Liu Y, Zeng X, Hu X, Hu J, Zhang X. Two-dimensional nanomaterials for photocatalytic water disinfection: recent progress and future challenges. *Journal of Chemical Technology & Biotechnology* 2019; **94**: 22–37.

23. Zhang X, Yuan X, Jiang L *et al*. Powerful combination of 2D g-C3N4 and 2D nanomaterials for photocatalysis: Recent advances. *Chemical Engineering Journal* 2020; **390**: 124475.
24. Lai C, An N, Li B *et al*. Future roadmap on nonmetal-based 2D ultrathin nanomaterials for photocatalysis. *Chemical Engineering Journal* 2021; **406**: 126780.
25. Zhong Y, Peng C, He Z *et al*. Interface engineering of heterojunction photocatalysts based on 1D nanomaterials. *Catalysis Science & Technology* 2021; **11**: 27–42.
26. Zhong S, Wang H, Wen H, Li J. The total factor productivity index of science and technology innovations in the coastal regions of China between 2006 and 2016. *Environmental Science and Pollution Research* 2021; **28**: 40555–67.
27. Wu M, Zhao M, Wu Z. Evaluation of development level and economic contribution ratio of science and technology innovation in eastern China. *Technology in Society* 2019; **59**: 101194.
28. Tang L. Five ways China must cultivate research integrity. *Nature* 2019; **575**: 589–91.
29. Reeb DM, Zhao W. Patents Do Not Measure Innovation Success. *Critical Finance Review* 2020; **9**: 157–99.
30. Crass D, Garcia Valero F, Pitton F, Rammer C. Protecting Innovation Through Patents and Trade Secrets: Evidence for Firms with a Single Innovation. *International Journal of the Economics of Business* 2019; **26**: 117–56.
31. Glaeser S. The effects of proprietary information on corporate disclosure and transparency: Evidence from trade secrets. *Journal of Accounting and Economics* 2018; **66**: 163–93.
32. Morikawa M. Innovation in the service sector and the role of patents and trade secrets: Evidence from Japanese firms. *Journal of the Japanese and International Economies* 2019; **51**: 43–51.
33. Brem A, Nylund PA, Hitchen EL. Open innovation and intellectual property rights. *Management Decision* 2017; **55**: 1285–306.
34. Contigiani A, Hsu DH, Barankay I. Trade secrets and innovation: Evidence from the "inevitable disclosure" doctrine. *Strategic Management Journal* 2018; **39**: 2921–42.
35. Holgersson M, Wallin MW. The patent management trichotomy: patenting, publishing, and secrecy. *Management Decision* 2017; **55**: 1087–99.